T0258785

CONTINUOUS
IMPROVEMENT

SHINGO MODEL SERIES

Discover Excellence: An Overview of the Shingo Model and Its Guiding Principles
Edited by Gerhard Plenert (2017)

Enterprise Alignment and Results: Thinking Systemically and Creating Constancy of Purpose and Value for the Customer
Edited by Chris Butterworth (2019)

Continuous Improvement: Seek Perfection, Embrace Scientific Thinking, Focus on Process, Assure Quality at the Source, and Improve Flow & Pull
Edited by Larry Anderson, Dan Fleming, Bruce Hamilton, and Pat Wardwell (2021)

CONTINUOUS
IMPROVEMENT

Seek Perfection, Embrace Scientific Thinking, Focus on Process, Assure Quality at the Source, and Improve Flow & Pull

Edited by

Larry Anderson, Dan Fleming, Bruce Hamilton, and Pat Wardwell

Routledge
Taylor & Francis Group

A PRODUCTIVITY PRESS BOOK

First published 2021
by Routledge
600 Broken Sound Parkway #300, Boca Raton FL, 33487

and by Routledge
2 Park Square, Milton Park, Abingdon, Oxon, OX14 4RN

Routledge is an imprint of the Taylor & Francis Group, an informa business

ISBN: 9781032105536 (hbk)
ISBN: 9781032107158 (pbk)
ISBN: 9781003216698 (ebk)

DOI: 10.4324/9781003216698

Typeset in Minion
by Apex CoVantage, LLC

Dedication

This book is dedicated to continuous improvement change agents from all corners and all levels of an organization. As the Shingo Institute wisely notes, "the transformation from traditional philosophy and practices to organizational excellence does not occur without the courage, creativity, and persistence of everyone in the organization—from executives to managers to team members on the frontline."

To the team members who have been transformed from doers to thinker-doers, we salute you. You are the early adopters who risk reprisal from doubters during the earliest stages of transformation. But we recognize also, as Shigeo Shingo noted, that people learn at different rates and that the greatest creative power is unleashed as more and more team members join the army of problem solvers. GBMP's slogan "Everybody Everyday" describes the ideal condition where improvement becomes a part of *every* person's daily routine.

To the managers who are transformed from isolated keepers of the status quo to aligned and passionate builders of effective systems for change (the means to improvement), we also dedicate this book. Here, too, there are early leaders whose insight and organizational savvy enable them to press the envelope of change. We applaud them. But, without the eventual engagement of *all* managers, continuous improvement will be localized at best, the early leaders martyred for their good works. Of all employees, managers and supervisors are challenged the most to embrace new roles and new behaviors as coaches and mentors. We have observed in teaching the Shingo Institute's workshops that, for this group, there is as much *un-learning* as there is *learning*. The bridge to sustainable improvement is built on that willingness to change.

Finally, we dedicate continuous improvement to senior executives, even up to directors, whose passionate leadership and vision is essential to create a workplace environment favorable to the use of improvement systems and tools. Shingo referred to this quality as *volition*, or impassioned will that aligns all team members to a new, radically different philosophy. As keepers and communicators of the Fundamental Principles, executives hold the keys to an organizational culture that can make improvement *continuous*.

Organizational excellence, as its name implies, requires the combined energies of the whole enterprise—executives, managers, and team members—each playing an essential complementary role to continuous improvement. We commend each of these change leaders for their "many small changes for the better" that collectively create better organizations and a better world.

Contents

Acknowledgments

We would like to thank Ken Snyder and his staff at the Shingo Institute for their patience with our starts and stops as we forged through the five fundamental principles of continuous improvement. We also want to acknowledge the participants from Shingo Institute workshops. They have shared insights and given us the customer viewpoint regarding the *Shingo Model* and its principles. We hope the amplification we have provided in this text will do justice to their thoughts and will support learning for all participants in the Institute's workshops. We are further indebted to Shingo Prize–recipient companies who embody and exemplify the culture and behaviors of enterprise excellence for all to see. Also, a sincere thank you to Utah State University and the Jon M. Huntsman School of Business. Since 1988, they have steadfastly supported the ideals of Dr. Shigeo Shingo by continuously supporting the Shingo Institute, the *Shingo Model*, the Shingo Prize, and the Institute's workshops. Finally, we offer our gratitude to Dr. Shigeo Shingo, a visionary hero who originally introduced the world to the principles, systems, and tools of continuous improvement and enterprise excellence.

In collaborating to produce this book, each of its four editors has at times shared their personal experiences as a student, teacher, Prize examiner, and consultant. This has generated many insightful first-person examples. We hope these vignettes will add more depth to the human side of operational excellence across many industries and settings. Our objective has been to support and amplify the personalized workshops that are given by registered Shingo Affiliates around the world.

Larry Anderson
Dan Fleming
Bruce Hamilton
Pat Wardwell

Acknowledgments

Introduction

In 1988, a Japanese industrial engineering consultant and author, Shigeo Shingo, bestowed his name on the "North American Shingo Prizes for Excellence in Manufacturing." While recognized for his genius by only a few individuals in the West, Shingo was highly regarded in Japan as a co-creator of the concepts, tools, and philosophy of the Toyota Production System (TPS). He was also the author of seventeen books on the subject.

Vernon Buhler, a director of Utah State University's (USU) Partners in Business program, was an early advocate of Shingo's teachings. It was Buhler who persuaded Shingo to accept an honorary doctorate in 1988 and to add his name to the Prize for Excellence in Manufacturing.

Shingo wanted the prizes, which were administered by the Jon M. Huntsman School of Business at USU, to be awarded each year to organizations and academics whose work exemplified the best of Shingo's teachings. The prizes were to be awarded in three categories: 1) large businesses of more than 500 employees, 2) small businesses of 500 or fewer employees, and 3) academics who made scholarly contributions to the body of knowledge surrounding Shingo's work.

The mission of the Shingo Prize was, as it is today, to recognize successful implementation of Shingo's ideas as examples of best practice for others to follow. Shingo wanted "to give back to North America" for what he himself had learned from his "teacher's teachers." These included Frank and Lillian Gilbreth, William Taylor, and Henry Ford.

Shigeo Shingo receives an honorary doctorate at Utah State University, 1988.

DOI: 10.4324/9781003216698-1

With a $50,000 donation from Norman Bodek, founder and former president of Productivity Press, and generous support from USU, the fledgling Shingo Prize presented its first award in 1989 at the 14th Annual Partners in Business Conference in Logan, Utah. By that time, several of Shingo's books had been translated into English from Japanese. This afforded organizations throughout the world the benefit of his incredible tools, such as SMED (single-minute exchange of dies) and poka-yoke (mistake-proofing). Perhaps even more valuable in Shingo's teachings were his observations on human nature and development, although the latter points were largely overlooked in favor of his tools in the early days of the Shingo Prize.

By 2008, Shingo's work and the significance of TPS beyond manufacturing became apparent to the Shingo Prize administrators. They expanded the scope of the Shingo Prize beyond North America and also made the Prize available to participants from non-manufacturing entities. The rebranded "Shingo Prize for Operational Excellence" was adopted. It included two additional significant but lesser levels of award: the silver and bronze medallions. Around this time, Prize administrators recognized the need and responsibility to provide a deeper understanding of the conceptual and philosophical foundations of Shingo's tools and methods, the *know-why* behind the *know-how*, as Shingo described them in his teachings.

HOW TO BEST USE THIS BOOK

What we refer to in the following chapters as the *Shingo Model*, the ten Guiding Principles, and the Three Insights of Organizational Excellence are the product of more than a decade of curriculum development by scholars and thought leaders to articulate a more robust explanation of Shingo's work. This explanation grounds the technical science of continuous improvement with a powerful social science that focuses on people development, which creates the opportunity for improvement to be truly continuous.

The whole point of the Shingo Institute's workshops, and of this book, is to impart a deeper understanding of the social/cultural side of organizational excellence. The technical part of organizational excellence—the tools—is necessary, but they are not sufficient to sustain the improvement engine. Without a nurturing and supportive culture that are guided by principles, the improvement process will not be continuous.

This text is primarily devoted to five of the ten Guiding Principles, collectively gathered under the heading Continuous Improvement. We specifically focus on 1) Seek Perfection, 2) Embrace Scientific Thinking, 3) Focus on Process, 4) Assure Quality at the Source, and 5) Improve Flow & Pull.

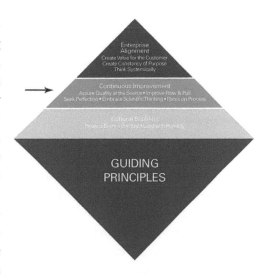

Naturally, the interconnection with the other five principles will necessarily require their occasional reference as well. Afterall, the *Shingo Model,* its *Guiding Principles,* and the three *Insights* are best understood as a whole.

Understanding also requires practice. As such, each chapter includes Reader Challenges that will engage you personally and help you to create the necessary reflection to make the *Shingo Model* a framework for you. Because the *Shingo Model* is a lens that is universal in its influence, it can be uniquely applied to you in context of your industry, discipline, and position of responsibility.

We will also attempt to shed light on how the principles and supporting concepts can be manifested in practice—not just understood in theory—using examples, questions, and stories from our own Lean journeys. Sometimes this will be done through examples that illustrate good results. Sometimes we will provide contrast by sharing anecdotes that relate to the misuse or misunderstanding of improvement principles and/or supporting concepts or tools. Our intent in each of the subsequent chapters is to enhance your comprehension of one or more aspects of the Continuous Improvement dimension and to increase your understanding of how this dimension interrelates with and complements others in the *Shingo Model.*

As noted previously, you will find Reader Challenges built into some chapters to encourage you to immediately apply what you have read within the context of your own organization. We believe tacit learning is critical to deepening your continuous improvement knowledge. This type of hands-on practice is necessary to understand the interrelatedness of principles, systems, and tools and is also critical in building confidence in your personal leadership skills.

It is our hope that the material presented here will inspire you to deepen your understanding of continuous improvement and to undertake further learning and practice within your own organization.

THE CONNECTIONS IN THE *SHINGO MODEL*

A brief word on the connecting arrows of the *Shingo Model* would be appropriate here. The third Insight of Organizational Excellence is "Principles Inform Ideal Behavior." The verb *inform* is defined as "giving an essential quality to." Thus, principles give an essential quality to ideal behavior. The *Shingo Model* provides guidance on how we accomplish this.

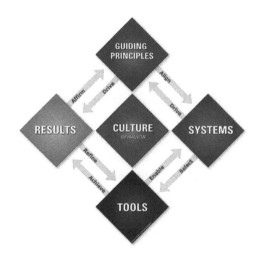

Starting with Guiding Principles and progressing in a clockwise direction, we *align* the systems we develop. To develop the ideal behavior culture (found at the center of the model), every system we develop aligns with the guiding principles. Continuing clockwise, the systems we develop determine which tools we *select* to implement the systems. Tools are the things we use every day to realize our results. Thus, we apply tools to *achieve* our business results. Finally, with results available, we can *affirm* that we are living the guiding principles. But how do we assure, at the outset, that our results will not indicate we have gone astray?

Now look at the *Model,* again starting with Guiding Principles, and proceed *counter*clockwise. The principles *drive* the results. In other words, in an ideal behavior culture, only certain results will be acceptable. Those are the ones we strive to achieve. Such results *refine* the understanding of the tools we should be using. These tools, since we use them every day, *enable* the systems to succeed in contributing to the ideal behavior culture. Finally, the systems, with reference to the second insight, Purpose and Systems Drive Behavior, *drive* living by the guiding principles.

Three Insights of Organizational Excellence
1. Ideal results require ideal behavior.
2. Purpose and systems drive behavior.
3. Principles inform ideal behavior.

Utilizing the *Shingo Model* ultimately gets us to the first insight, Ideal Results Require Ideal Behavior. If we create an ideal behavior culture, we will achieve ideal results.

1

Organizational Excellence and the Shingo Institute

Too many organizations are failing to be competitive, not because they cannot solve problems, but because they cannot sustain the solution. They haven't realized that tradition supersedes tools, no matter how good they are. Success requires a sustainable shift in behaviors and culture, and that needs to be driven by a shift in the systems that motivate those behaviors.[1]

—Gerhard Plenert

For every organization, the answer to the question, "What does excellence mean to your organization?" will always be different. No two organizations will respond the same way. Following are some of the key definitions of organizational excellence that the Shingo Institute has learned from organizations across the world.

Organizational excellence is:

1. Instilling purposeful change to mitigate the root cause of performance problems.
2. Improving the work. It is not just about the processes. It is about engaging the whole workforce and making a better company.
3. Creating a great environment for the people, looking for the same objectives, and having fun everywhere.
4. A status in which an organization has not only achieved financial or market results, but also transcended to a different level where respectful people, culture, and principles are key factors in a strategy to sustain growth over generations.

---

[1] Gerhard Plenert, *Discover Excellence: An Overview of the Shingo Model and Its Guiding Principles* (Boca Raton, FL: CRC Press, 2018), 1.

DOI: 10.4324/9781003216698-2

5. Understanding what customer needs are.
6. Trying to optimize the value for customers.
7. Being never satisfied. Constantly moving forward, constantly thinking. An organization where it is safe to challenge everything.
8. A focus on culture and behaviors.
9. One that enhances culture because it brings focus on the customer.

BACK TO BASICS

The term *Lean* was first introduced in 1990 in the book *The Machine That Changed the World: The Story of Lean Production*. In it, the authors, James Womack, Daniel Jones, and Daniel Roos, describe Lean as manufacturing systems that are based on the principles employed in the Toyota Production System (TPS). They wrote:

> *Lean . . . is "lean" because it uses less of everything compared with mass production—half the human effort in the factory, half the manufacturing space, half the investment in tools, half the engineering hours to develop a new product in half the time. Also, it requires keeping far less than half the inventory on site, results in many fewer defects, and produces a greater and ever-growing variety of products.*[2]

In the intervening years, the philosophy of Lean has gone through numerous iterations. It stresses the maximization of customer value while simultaneously minimizing waste. The goal of Lean is to create increased value for customers while simultaneously utilizing fewer resources. Countless organizations have, at one time or another, begun a Lean journey or implemented an improvement initiative of some sort. At the foundation of these initiatives are a plethora of tools (over 100) that seem to promise exciting new results. They are utilized to optimize the flow of products and services throughout an entire value stream as they horizontally flow through an organization.

While many organizations may initially see significant improvements, far too many of these initiatives meet disappointing ends. Leaders quickly find that tools such as Six Sigma, SMED, 5S (sort, set in order, sweep, standardize, and sustain), and just-in-time (JIT) are not independently capable of effecting lasting change.

---

[2] J. P. Womack et al., *The Machine That Changed the World: The Story of Lean Production—Toyota's Secret Weapon in the Global Car Wars That Is Now Revolutionizing World Industry* (New York, NY: Simon & Schuster, Inc., 1990), 14.

THE SHINGO INSTITUTE

The Shingo Institute has assessed organizations in various industries around the world. The Institute has seen firsthand how some organizations have been able to sustain their improvement results, while far too many have experienced such a decline. In fact, initially, the Shingo Prize focused on tools and systems and how those tools and systems drive results. The Prize was originally given out on the basis of these results.

But when far too many Shingo Prize–recipient organizations reverted to their old ways, the Shingo Institute realized there was a big piece missing in its earlier model of organizational excellence based only on systems, tools, and short-term results. So, the Shingo Institute set out to determine the key difference between short-lived successes and sustainable results. Over time, the Institute discovered a common theme: the difference between sustainable and unsustainable effort is centered on the ability of an organization to ingrain into its culture timeless and universal principles, rather than rely on the superficial implementation of tools and programs. This is because principles help people understand the "why" behind the "how" and the "what." Sustainable results depend upon the degree to which an organization's culture is aligned to specific, guiding principles rather than depending solely on tools, programs, or initiatives.

The Shingo Institute discovered that what was lacking was sustained superior performance, a sustained culture of excellence and innovation, and a sustained environment for social and ecological leadership. To really make progress in a journey to organizational excellence, we must have long-term sustainability. Change could no longer be something that happened once a year during a Lean event. Instead, organizations need to constantly look for improvement opportunities.

THE *SHINGO MODEL* AND THE SHINGO PRIZE

To best illustrate its new findings, the Shingo Institute developed the *Shingo Model*™, the accompanying *Shingo Guiding Principles*, and the *Three Insights of Organizational Excellence*™. The principles are timeless and universal. They apply to all cultures and they do not change over time. They provide a solid foundation for developing a roadmap to excellence.

Now, the Shingo Prize is awarded to organizations that have robust key systems driving behavior closer to ideal, as informed by the principles of organizational excellence, and measured by strong key performance indicator (KPI) and key behavioral indicator (KBI) trends and levels. Shingo Prize recipients show the greatest potential for sustainability as measured by the frequency, intensity, duration, scope, and

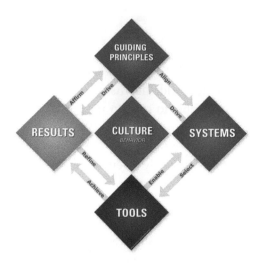

role of the behaviors evident in the organizational culture. The Shingo Prize has become the global standard for organizational excellence. As an effective way to benchmark progress toward excellence, organizations throughout the world may apply and challenge for the prize. Recipients receiving this recognition fall into three categories: Shingo Prize, the Shingo Silver Medallion, and the Shingo Bronze Medallion.

Most organizations do not wait until they believe they might qualify for the Shingo Prize to challenge for it. In fact, many organizations do not intend to ever challenge for the Shingo Prize. But they use the *Shingo Model* and the assessment process to measure themselves as they work toward the highest standard of excellence in the world. They use the guidelines to direct them, to inspire them, and to hold themselves responsible.

Over the years, the Shingo Institute's scope has expanded to include various educational offerings, a focus on research, and a growing international network of Shingo Institute Licensed Affiliates. The *Shingo Model* is the primary subject of the Institute's popular workshops and publications. These materials have been developed to share throughout the world so organizations can learn how to create a sustainable cultural shift, which will ultimately lead to organizational excellence.

Similarly, volunteer Shingo examiners, who are international experts in all aspects of organizational excellence, focus on determining the degree to which the *Shingo Guiding Principles* are evident in the behavior of every team member in an organization. They observe behavior and determine the frequency, intensity, duration, scope, and role of the desired principle-based behaviors. They observe the degree to which leaders are focused on principles and culture, and the degree to which managers are focused on aligning systems to drive ideal behaviors at all levels.

As part of its educational offerings, the Shingo Institute offers a series of six workshops that are designed to help participants understand the *Shingo Model*, its *Guiding Principles*, and its *Insights*. Ultimately, these workshops help participants strive for excellence within their respective organizations. Each of the workshops is described below.

DISCOVER EXCELLENCE

 This foundational, two-day workshop introduces the *Shingo Model*, the *Shingo Guiding Principles*, and the *Three Insights to Organizational Excellence*. With active discussions and on-site learning at a host organization, this program is a highly interactive experience. It is designed to make learning meaningful and immediately applicable as participants discover how to release the latent potential in an organization to achieve organizational excellence. It provides the basic understanding needed in all Shingo workshops; therefore, it is a prerequisite to all the other Shingo workshops. During this workshop, participants will learn and understand the *Shingo Model*, discover the *Three Insights of Organizational Excellence*, and explore how the *Shingo Guiding Principles* inform ideal behaviors that ultimately lead to sustainable results. They will also understand the behavioral assessment process through an interactive case study and on-site learning.

SYSTEMS DESIGN

 This two-day workshop integrates classroom and on-site experiences at a host facility to build upon the knowledge and experience gained in the DISCOVER workshop, and focuses on the Systems and Tools diamonds in the *Shingo Model*. It begins by explaining that all work in an organization is the outcome of a system. Systems must be designed to create a specific

end objective; otherwise, they evolve on their own. Systems drive the behavior of people, and variation in behavior leads to variation in results. Organizational excellence requires well-designed systems to drive ideal behaviors that are required to produce sustainable results. During this workshop, participants will discover three types of essential systems and explore five required communication tools for each system. They will also learn how to create and use system maps and understand system standard work and how it drives improvement.

CULTURAL ENABLERS

This two-day workshop integrates classroom and on-site experiences at a host facility to build upon the knowledge and experience gained in the DISCOVER and SYSTEMS workshops. It takes participants deeper into the *Shingo Model* by focusing on the principles identified in the Cultural Enablers dimension: Respect Every Individual and Lead with Humility.

Cultural Enablers principles make it possible for people in an organization to engage in the transformation journey, progress in their understanding, and build a culture of enterprise excellence. Organizational excellence cannot be achieved through top-down directives or piecemeal implementation of tools. It requires a widespread organizational commitment. The CULTURAL ENABLERS workshop will help participants define ideal behaviors and the systems that drive those behaviors using behavioral benchmarks.

CONTINUOUS IMPROVEMENT

This two- or three-day workshop integrates classroom and on-site experiences at a host facility to build upon the knowledge and experience gained in the DISCOVER and SYSTEMS workshops. It begins by teaching participants how to clearly define value through the eyes of their customers. It continues the discussion about ideal behaviors, fundamental purpose, and behavioral benchmarks and takes participants deeper into the *Shingo Model* by focusing on the principles identified in the Continuous Improvement dimension: Seek Perfection, Embrace Scientific Thinking, Focus on Process, Assure Quality at the Source, and Improve Flow & Pull. The CONTINUOUS IMPROVEMENT workshop will deepen participants' understanding of the relationship between behaviors, systems, and principles and how they drive results.

ENTERPRISE ALIGNMENT

This two-day workshop integrates classroom and on-site experiences at a host facility to build upon the knowledge and experience gained in the DISCOVER and SYSTEMS workshops. It takes participants deeper into the *Shingo Model* by focusing on the principles identified in the Enterprise Alignment dimension: Think Systemically, Create Constancy of Purpose, and Create Value for the Customer.

To succeed, organizations must develop management systems that align work and behaviors with principles and direction in ways that are simple, comprehensive, actionable, and standardized. Organizations must get results, and creating value for customers is ultimately accomplished through the effective alignment of every value stream in an organization. The ENTERPRISE ALIGNMENT workshop continues the discussion around defining ideal behaviors and the systems that drive them.

BUILD EXCELLENCE

This two-day capstone workshop integrates classroom and on-site experiences at a host facility to solidify the knowledge and experience gained from the previous five Shingo workshops. The BUILD EXCELLENCE workshop demonstrates the integrated execution of systems that drive behavior toward the ideal as informed by the principles in the *Shingo Model*. The workshop helps develop a structured approach to execute a cultural transformation. It builds upon a foundation of principles, using tools that already exist within many organizations. Participants will learn how to build systems that drive behavior that will consistently deliver desired results.

In this final Shingo workshop, participants discover how the *Shingo Model* strengthens the execution of their strategy. They learn how to shift the culture, guided by organizational purpose, to the next level. They use Go & Observe to understand the practical application of the *Shingo Model*. Finally, they accelerate the cultural transformation by applying the learnings gained from the *Shingo Model* into a PDCA cycle.

THE SHINGO INSTITUTE'S BOOK SERIES

In conjunction with the Shingo Workshop series, the Shingo Institute has set out to publish six books that are specifically focused on the primary components of the *Shingo Model* and its guiding principles. This book, *Continuous Improvement*, is one of the books in the series. Two more, *Discover Excellence* and *Enterprise Alignment and Results*, have been published and others are on the horizon.

In all of these efforts, the focus at the Shingo Institute is unique in the world. Its work is the most rigorous way to determine if an organization is fundamentally improving over the long term. Its goal is to help every organization reach continuous improvement—wherever it may be along its path.

2

The Continuous Improvement Dimension

Improvement means the elimination of waste, and the most essential pre-condition for improvement is the proper pursuit of goals. We must not be mistaken, first of all, about what improvement means. The four goals of improvement must be to make things easier, better, faster, and cheaper.[1]

—Shigeo Shingo

CONTINUOUS IMPROVEMENT IN A NUTSHELL

The *Shingo Model* has at its core the relentless quest to improve an organization. This is driven from a culture that nurtures, aligns, and drives behaviors in support of achieving desired results. Collectively, the *Shingo Model* helps us focus on *who, what, when, where, why,* and *how* organizations need to understand to pursue results that are both ideal and sustainable. Shingo said:

A key activity emphasized at Toyota is finding the true causes of problems or waste. We ask "Why?" again and again until the answer is found. Traditionally, 5W 1H means:

Who: subject of production
What: objects of production
When: time
Where: space

[1] Shigeo Shingo, *The Sayings of Shigeo Shingo*, trans. Andrew P. Dillon (Cambridge, MA: Productivity, 1987), 43.

DOI: 10.4324/9781003216698-3

Why: to find the cause for each of the above because they are all important factors in unravelling a problem
How: methods

At Toyota, the five Ws really mean five Whys—asking "Why?" five or
more times until the cause of a problem is discovered. For every factor—
what, who, where, when, and how—we ask "Why, why, why, why, why?"
Asking once is never enough. By asking "Why?" five times, how we should
solve the problem is also clarified.[2]

While the Cultural Enablers and Enterprise Alignment dimensions of the
Shingo Model address the *who* and *why* principles inherent in the journey, the Continuous Improvement dimension addresses the *what, when,
where*, and *how* principles. More specifically, Continuous Improvement
helps organizations understand the context, mindset, and tools that the
Shingo Model introduces so they can analyze and transform systems and
processes in the pursuit of excellence.

People who achieve the best long-term results in their organizations are
acutely aware that the study and practice of improvement, according to the
Shingo Model, involves both technical and social sciences—and the social
sciences carry a stronger weight, as we learn in the study of the Cultural
Enablers dimension. This is because all valuable work and improvement are
accomplished through people. People are what make the improvement culture sustainable over time. Ultimately, the quest is for excellent results, which
are generally measured by KPIs. But the social side of this model (culture
and behaviors) is what delivers those results. This is why the *Shingo Model*
places a strong emphasis on ideal behaviors and promotes a shift to include
KBIs as leading indicators for any organization on the Shingo journey.

Of course, without a correspondingly strong understanding of the technical science of improvement, namely, how to systematically analyze,
change, and improve systems and processes, organizations will struggle to
achieve the full benefits of the *Shingo Model*.

For example, Shingo said that the goals of improvement are to make things
easier, better, faster, and cheaper. But if you aren't able to answer questions
such as, "Better for whom?" or "What is meant by faster?" or "Cheaper
in what regard?" your improvements may be directionally incorrect, limited, or unsustainable. The technical side of improvement is foremost in

---

[2] Shigeo Shingo, *A Study of the Toyota Production System*, trans. Andrew P. Dillon (Portland, OR:
Productivity, 1989), 82.

the Continuous Improvement dimension of the *Shingo Model*. It requires a deep understanding of the five principles and the many Lean supporting concepts and tools that can be used to develop and reinforce them.

In plain terms, the principles in this dimension are aimed at systematically improving the flow of value to customers through an iterative process of identifying and then by reducing or eliminating all the things that get in the way of flowing value (e.g., eliminating waste). The idea is that organizations have a responsibility to pursue perfection (i.e., the principle Seek Perfection) because it is through the act of making improvement and gaining tacit learning that the correct mindset and culture of continuous improvement are developed. The *Shingo Model* recognizes that the lofty goal of perfection is not likely to be achieved. And that's okay. What is *not* okay is to stop continuously advancing toward it.

The other four principles within the Continuous Improvement dimension go further in providing guidance and specificity in terms of what it means to improve under the *Shingo Model*. If we start with the principle Improve Flow & Pull, we find that value for customers is highest when it is created in response to real demand and at a continuous and uninterrupted flow, with one-piece flow at the customer's desired rate of demand being the ideal we must seek. Accepting this principle, it becomes clear why the Quality at the Source principle is critical. When quality issues occur, customer flow is negatively impacted and value creation is corrupted. Conversely, when perfect quality is achieved by completing every element of work correctly the first time, customer value can flow.

The remaining two principles address the means by which we optimize flow and pull and ensure quality at the source. By Focusing on Process, we send the message that all problems related to flow, pull, and quality are rooted in an imperfect process and not in the people involved. Therefore, we need an accurate and up-to-date understanding of organizational processes to recognize problems and uncover opportunities. The best way to solve problems and make improvements is by Embracing Scientific Thinking. This calls for repeated cycles of structured experimentation, direct observation, and tacit learning that come from embracing new ideas, failures, and the constant refinement and challenging of current process reality.

As we will point out many times in the pages ahead, all component parts and principles related to the *Shingo Model* are interdependent and do not stand on their own. Although this book is focused on the Continuous Improvement dimension, we encourage you to study the interrelationship of all three dimensions. Like an ecosystem, you will discover strong dependencies between the people, process, and purpose principles and their supporting concepts. Changes or adjustment within any one dimension will impact the others.

THE PRINCIPLES

Listed below are the Shingo Institute's definitions of the five principles in the Continuous Improvement Dimension. This section is followed by a brief discussion of the nine supporting concepts that we hope will bring the principles and concepts to life. In later chapters, we will provide deeper insight into the principles and show how specific supporting concepts reinforce each one.

Enterprise
Alignment
Create Value for the Customer
Create Constancy of Purpose
Think Systemically

Continuous Improvement
Assure Quality at the Source • Improve Flow & Pull
Seek Perfection • Embrace Scientific Thinking • Focus on Process

Cultural Enablers
Respect Every Individual • Lead with Humility

GUIDING
PRINCIPLES

> **Continuous improvement is not about the things you do well—that's the work. Continuous improvement is about removing the things that get in the way of the work. The headaches, things that slow you down; that's what continuous improvement is about.**[3]
>
> —*Bruce Hamilton*

Seek Perfection: *Perfection is an aspiration not likely to be achieved, but the pursuit of perfection creates a mindset and culture of continuous improvement. What is possible is only limited by the paradigms through which we see and understand the organization's current reality.*[4]

Embrace Scientific Thinking: *Innovation and improvement are the consequence of repeated cycles of experimentation, direct observation, and*

---

[3] Bruce Hamilton, *Toast Kaizen: An Introduction to Continuous Improvement & Lean Principles* (Boston, MA: GBMP, 2005), DVD.

[4] Shingo Institute, *The Shingo Model,* version 14.5 (Logan: Utah State University, 2020), 22.

learning. A relentless and systematic exploration of new ideas, including failures, enables us to constantly refine our understanding of reality.[5]

Focus on Process: *All outcomes are the consequence of a process. It is nearly impossible for even good people to consistently produce ideal results with poor processes. It is human nature to blame the people involved when something goes wrong or when the resulting product or service is less than ideal. But in reality, an issue is usually rooted in an imperfect process, not in the people involved.*[6]

Assure Quality at the Source: *Perfect quality can only be achieved when every element of work is done right the first time. If a defect occurs in a product or service, it must be detected and corrected at the time it is created.*[7]

Improve Flow & Pull: *Value for customers is highest when it is created in response to real demand and at a continuous and uninterrupted flow. Although one-piece flow is the ideal, demand is distorted between and within organizations. Waste is anything that disrupts the continuous flow of value.*[8]

THE SUPPORTING CONCEPTS

The supporting concepts of the Continuous Improvement principles, like the *Shingo Model* principles, work in conjunction with one another to create the opportunity for excellent results. They are presented here as they appear in the *Shingo Model* booklet, along with a brief discussion.

Stabilize Processes: *Stability in processes is the bedrock foundation of any improvement system. It creates consistency and repeatability and is the basis for problem identification. Nearly all continuous improvement principles rely on stability because it is the precursor to achieving flow. Many of the rationalizations for waste are based on the instability of processes, as if they are beyond our control. Instead, organizations should apply the basic tools available to reduce or eliminate instability and thus create processes that help to identify and eliminate waste.*[9]

---

[5] Ibid., 22.
[6] Ibid., 22–23.
[7] Ibid., 23.
[8] Ibid.
[9] Ibid.

The word *stabilize* conveys the action of "holding steady," which indicates minimal variation from cycle to cycle. We have said to team members, "If you want to make it difficult to improve, perform process steps a different way each time." Seeing an entire flow of value and identifying waste requires iterations of observation. The more variation and the less steadiness that exists from cycle to cycle, the more difficult it is to see patterns. All Lean tools can, in some way, contribute to process stability. Production processes are an obvious place to develop standard conditions, but they are not sufficient, in and of themselves, to create overall process stability. There must also be minimal variation in people's understanding of the purpose, vision, and mission of an organization. It requires consistent policy deployment at all levels of an organization to result in minimal variation in alignment.

Standard Work: *While stability is a necessary precondition for creating flow and improvement, creating standard work builds control into the process itself. Standard work is the supporting principle behind maintaining improvement rather than springing back to preceding practices and results. Standard work also eliminates the need to control operations through cost standards, production targets, or other traditional supervisory methods. When standard work is in place, the work itself serves as the management control mechanism, and it becomes easier and faster to check the status of processes and operations. Supervisors are freer to work on other tasks when there is less need to monitor and control the work process.*[10]

This concept conveys the idea of "bringing into conformity with a standard especially in order to assure consistency and regularity." Applying Shingo's definition of process, which is that processes transform materials into products, the implication is that we establish expectations for the transformation. This includes not only the specifics of Shingo's operations, or "the way the work is done," but also the overall flow of value across the organization. Historical observation of organizations indicates that, if there are standards, they tend to focus on the operations. There is rather limited consideration for determining how the process should flow. Focus on process would indicate that there are standards for how long a component should be at a location, whether related to production or movement. Processes should be tracked and analyzed while considering the established standard, thus allowing organizations to see anomalies in flow.

---

[10] Ibid., 24.

Go & Observe: *Direct observation is a supporting principle tied to scientific thinking. It is, in fact, the first step of the scientific method. Observation is necessary to truly understand the process or phenomenon being studied. All too frequently, perceptions, past experiences, instincts, and inaccurate standards are misconstrued as reality. Through direct observation, reality can be seen, confirmed, and established.*[11]

One cannot understand what is going on in a process, or operations for that matter, without seeing it for oneself. We have observed executives and managers who believe that reports can indicate to them the status of a process. But any report is backward looking. There is always some delay between the fact and the report, which separates the creation point of a problem from the point of discovery. This means that leadership behavior, in particular, must include being in the production or support work areas. It also means that team members must be capable of analyzing their work environment to determine how it impacts the flow of value. Organizations practicing this concept will have executives systematically visit all work areas, managers present as a matter of practice in the work area, and team members being provided mechanisms to analyze work and record their continuous observations.

Focus on Process: *Improve Flow & Pull combined with Focus on Process necessitates defining value streams and focusing on them. A value stream is the collection of all the necessary steps required to deliver value to the customer. Defining what customers value is an essential step in focusing on the value stream. Clearly understanding the entire value stream, however, is the only way for an organization to improve the value delivered and/or to improve the process by which it is delivered.*[12]

The value stream includes all the activities required to bring a product of service to the customer. From inception to delivery, each of the elements of the flow of value must be considered. If an organization is exhibiting a focus on process, it will naturally take the perspective of an entire value stream. Value stream perspective allows practitioners to achieve the focus that Shingo suggests: improve the process (meaning the flow of value through all operations), then improve the transformation operations. Many times, though, we see the opposite. The organization is focused on

---

[11] Ibid.
[12] Ibid.

improving work elements without considering the impact on the overall flow of value. Value stream analysis identifies the disruptions in the flow of value by identifying the factors, such as inventory stagnation, that contribute to process and batch delay. An organization that is practicing a focus on process will show value stream thinking.

Keep It Simple and Visual: *In society today, we frequently see a bias toward complex solutions as well as a premium paid to those who seem to manage complexity well. But usually, better results at a lower cost can be achieved by simplification. Dr. Shingo's life work in mistake proofing is centered on this principle. Much of waste is the result of information deficits. Making information visual, when combined with simplification, solves the information deficits.*[13]

A behavioral check of this concept is if an outside observer can determine in seconds the current status of a process, including the overall flow of value and the status of the contributing operations. If this is not the observed condition, there can be significant consequences. First, since executives are charged with looking at the whole, there is a relationship between the time required to see the whole and the level of sustainable commitment executives will exhibit. If the process is difficult to see, executives, with their myriad obligations, will find it difficult to spend the time necessary to fully understand the process.

If this is the condition, then, over time, executive presence where the work is done will diminish. We often talk to teams about their visual performance boards in the context of wasting executives' time. We tell them that if executives feel like visits to review the boards are a waste of their time, they will stop coming. This builds on another Shingo principle—respect for every individual. A team should determine if the way they present the board's information is being respectful of the executive's time.

Performance boards themselves might also be considered backward looking, depending on their structure and what is being reported. Updating cycles for process and operation metrics must be consistent with immediate status identification, and a deep understanding of workplace organization must be evident. The essence of this concept is incorporating the elements of 5S and visual management in every activity. Not only should elements be applied to the physical workplace, but also instructions, drawings, forms, and all other communication methods should be equally structured to be simple and visual.

---

[13] Ibid., 25.

Identify and Eliminate Waste: *Identification and elimination of waste is a practical concept for making processes flow, thus it becomes a primary focus of continuous improvement. Waste is anything that slows or disrupts the continuous flow of value to customers. Waste elimination is a powerful supporting principle because it is easily understood. Focusing on the elimination of waste will consistently drive appropriate behavior, while the wrong focus can frequently become a barrier to improvement. In the end, identifying and eliminating waste will engage the entire organization in the continuous improvement effort.*[14]

Recognizing that a major portion of the disruption in process flow is between the operations, a focus on process is the optimum way to identify waste. The tool of value stream mapping enables us to see and quantify the impact of waste on flow of value. Elimination of waste can occur only when other interdependent concepts are acted upon. Focus on process will require that business personnel prioritize waste identification as a first order of business. Waste elimination, not improvement, should become the objective. In the process, improvement will result from ongoing elimination of waste. Visibility of waste is dependent on the success of other concepts. Remember Shingo's perspective on waste: improve processes before improving operations.

No Defects Passed Forward: *This concept is essential for organizational excellence from many different points of view. From an executive's perspective, it requires great courage to stop the process long enough to understand the root cause and to take countermeasures that prevent the process from reoccurring. This often means trading any short-term loss for substantial long-term gain. From a manager's perspective, systems must be in place to ensure that any result that varies from the standard, even slightly, creates an expectation of and support for immediate action. This is often called "swarming." From a team member's point of view, no defect passed forward requires a mindset of ownership and accountability. If standards are clearly defined, every person should know what good is. Executives and managers should role-model and then create the conditions for team members to develop the mindset of personal integrity. This means that no one would ever knowingly or willingly forward the outcome of their value contribution to someone else if it contained the slightest variation from the standard. This supporting concept feeds the mindset and tools of continuous improvement*

---

[14] Ibid.

and creates the conditions for seeking perfection. It is possible to achieve perfection in the application of this concept.[15]

This concept focuses on the accountability executives and management give team members to stop production, in whatever way necessary, so that problem solving can be conducted in real time and defects don't reach the customer. Real-time problem solving is the most effective way to resolve issues. We recognize that both problem and solution are in the workplace, but if this concept is not applied, the point of creation will begin to deviate from the point of discovery. An organization that does not provide mechanisms for stopping when a potential defect is identified is not showing respect for its people. When the point of creation of a defect is removed from the point of discovery, the natural question becomes, "How did it get this far?" Since the process itself does not care about results, often the natural tendency is to blame the people in the process. (After all, the process is just a set of machines, instructions, standard operating procedures [SOPs], etc.) Executives and management must provide good processes to avoid this tendency. One way to do this is to build autonomation (automation with a human touch) into processes. This will ensure that operations include inherent assurance of quality.

Integrate Improvement with Work: *As the migration toward a principle-based culture occurs, the activities and approaches for continuous improvement become part of the everyday work of every employee in an organization. Each person in an organization performs daily work. When improvement is integrated with work, each person naturally accepts responsibility for improvement of the daily work processes. Executives are responsible for improving strategy-setting processes or perhaps resource alignment and employee development processes. They are primarily responsible for deploying mission-critical strategy and metrics into the organization so that every person not only has a clear line of sight to what matters most but are also motivated by the mission in a way that creates a compelling case for improvement. Managers are responsible for improving quality systems, or performance development systems, or value stream flow. Team members are responsible for improving their cycle times, or quality of work, or yields. They become "scientists" who continually assess the current state of their processes and pursue a better, future state that will enhance value (or eliminate waste) and thus, pursue perfection. Integrating improvement with work is more*

---

[15] Ibid., 25–26.

than assigning responsibility; it entails the creation of standardized work that defines the necessary systems for improvement.[16]

As waste is reduced, time becomes more readily available so improvement activity can be built into the process instead of being added to the existing work. Waste quite often exposes itself through the behaviors of the people involved in the process, so we have an ongoing need to ask why they are doing what they are doing. Any of the wastes require a response by the process, and the only resource for the response is the people involved, so both good and bad behaviors are driven. The practical application of this concept is providing time in the workday to focus on improvement. This might begin as a commitment to dedicated improvement activities, kaizen events, or natural problem-solving teams, which should be considered batch improvement activities. As executives, managers, and team members perfect their ability to effectively utilize this time, their efforts should become more flow-related and improvement will be more real-time and spontaneous. In order to most effectively synthesize work and improvement, the interdependent components of the *Shingo Model* must also be applied.

Rely on Data and Facts: *Dr. Shingo emphasized the importance of being data-driven in the pursuit of continuous improvement. He frequently shared examples of specific situations where data was collected, but the data was incorrect or wasn't actually used in the improvement process. He was adamant that the understanding of the actual process be so thorough that when implementing a change in the process, the improvement, as evidenced by the data, could be predicted. Thus, reconciliation would be required between the predicted results and the actual results, making the improvement process truly data driven. Ultimately, when data is treated loosely or imprecisely, the tendency is to leave potential improvement on the table or, even worse, to not achieve any improvement at all.*[17]

One definition of the word *fact* is "something that has actual existence." The supporting concept, Rely on Data and Facts, requires all members of an organization to see, feel, touch, and hear the reality of the process. Shingo says, "The greatest waste is the waste we don't see."[18] He also says, "The most damaging kind of waste is the waste we don't recognize."[19]

---

[16] Ibid., 26–27.

[17] Ibid., 27–28.

[18] Shingo, *The Sayings of Shigeo Shingo*, 19.

[19] Shigeo Shingo, *The Shingo Production Management System,* trans. Andrew P. Dillon (Cambridge, MA: Productivity, 1992), 35.

The primary behavioral sign of this concept is how people talk about opportunities and issues. You would not hear conversations including indefinite phrases such as "I think" and "they said," for example. *Data*, likewise, is defined as "factual information (such as measurements or statistics) used as a basis for reasoning, discussion, or calculation." Data is the representation of facts in ways that others can see and understand. Executives and management must provide team members with simple, visual mechanisms to present the facts associated with their parts of the process. This primarily comes from establishing a set of standard conditions so that observation of anomalies can easily be made and recorded. As long as the process is working to standard, time can be spent on improving the flow of value to the customer. If a non-standard condition develops, it becomes obvious (it is a *fact*) and immediate problem solving can be initiated to return it to standard. When we set an expectation that facts are recorded, analysis to understand patterns and trends also becomes possible.

3

Seek Perfection

True North: What we should do, not what we can do.[1]

—Hajime Ohba

BUSINESS CASE FOR SEEK PERFECTION

The principle Seek Perfection is succinctly defined by the Shingo Institute: *Perfection is an aspiration not likely to be achieved, but the pursuit of perfection creates a mindset and culture of continuous improvement. The realization of what is possible is only limited by the paradigms through which one sees and understands the organization's current reality.*[2]

FUNDAMENTAL TRUTH

According to the Shingo Institute, there is also a fundamental truth behind the principle Seek Perfection: "People have an innate desire and ability to improve that is only limited by their expectations."[3] In other words, because we have an innate desire and ability to improve which is limited only by our expectations, we seek perfection.

---

[1] Toyota Production System Support Center (TSSC), joint training with GBMP, Westborough, MA, July 30, 2003.
[2] Shingo Institute, *The Shingo Model*, 22.
[3] Shingo Institute, *Cultural Enablers Workshop* (Logan, UT: Shingo Institute, November 2020).

DOI: 10.4324/9781003216698-4

Illustration of MacGregor's X and Y
Theory.

In his book, *Non-Stock Production*, Shingo added an important quali-
fication based on Douglas MacGregor's X and Y Theory.[4] Shingo suggested that workers who are treated by managers as selfish and lazy, for
example, will behave that way, while those for whom expectations are high
will also respond accordingly. In other words, while desire and ability are
innate human qualities, they will likely be either developed or suppressed,
depending on the culture of the organization. Because the search for perfection relies on a culture that favors the search, we'll first describe the
salient characteristics of a continuous improvement culture.

KEY SUPPORTING CONCEPTS

Two key supporting concepts, Integrate Improvement with Work and
Identify and Eliminate Waste, provide a deeper perspective on the mindset and behaviors needed to Seek Perfection.

Integrate Improvement with Work: *As the migration toward a
principle-based culture occurs, the activities and approaches for continuous improvement become part of the everyday work of every employee in
an organization.*[5] See the following discussion for further detail on this
supporting concept.

---

[4] Shigeo Shingo, *Non-Stock Production: The Shingo System for Continuous Improvement* (Cambridge,
MA: Productivity, 1988), 18.
[5] Shingo Institute, *The Shingo Model*, 26.

Identify and Eliminate Waste: *Identification and elimination of waste is a practical concept for making processes flow, thus it becomes a primary focus of continuous improvement.*[6] See the following discussion for further detail on this supporting concept.

SEEKING PERFECTION IS THE FLYWHEEL

Of the five principles associated with the Continuous Improvement dimension, Seek Perfection is the flywheel to endless daily improvement that brings goods and services to the customer. But the concept of perfection is elusive. It can be viewed as the *result* of our actions or as the *actions* themselves. This is underscored by the first of the Three Insights of Organizational Excellence: *Ideal results require ideal behavior.*[7]

For example, a perfect baseball game is one in which no opposing players reach first base for any reason. But the perfect game also connotes the human endeavor that leads to the result, in this case, the performance of the nine players making many great plays. The ideal itself is daunting. While every team aspires to a perfect game, there have only been twenty-three of them played in more than a century. The odds of achieving this lofty ideal are roughly one in fifty thousand. In professional football, the perfect result is defined in different terms: a season and post-season with no defeats and no ties. Every season for the past one hundred years, every team has aspired to this ideal, but it has been achieved only once.

So why seek perfection? Isn't it a bit impractical and perhaps even discouraging? Legendary football coach Vince Lombardi said, "If we chase perfection, we can catch excellence." In other words, the chase "creates the mindset and culture of continuous improvement." Ideal results and ideal behaviors exist in the context of the systems to which they are applied. While the paradigms for perfection in baseball or football are universal, the same cannot be said for all endeavors. In fact, the mindsets in most organizations today are often based upon outmoded paradigms that implicitly reject the concept of perfection.

Therefore, to understand and embrace this concept, we are encouraged by the Shingo Institute to "challenge our paradigms and expectations."[8]

---

[6] Ibid., 25.

[7] Ibid., 10.

[8] Shingo Institute, *Continuous Improvement Workshop* (Logan, UT: Shingo Institute, November 2020).

A deeper understanding of the principle requires close examination of the paradigms themselves, both old and new, as they are basic frameworks through which we organize our thinking and decision making. Old paradigms, referred by Shingo as "conceptual blind spots,"[9] can distort our interpretations of what we see and therefore limit our understanding of the world. New paradigms create both the vector and the creative force that define improvement and render it continuous. In this chapter, we'll dig a little deeper into the old and new paradigms and provide examples of tools and systems that drive each of them. In the process, we'll work to answer the following questions: What is perfection and is it relative to each organization or is it universal? Why is pursuing perfection necessary? Can we be excellent without being perfect? And, finally, who are the seekers of perfection?

ELABORATION ON BEHAVIORAL BENCHMARKS

The answers to these questions are best described in two Behavioral Benchmarks, which are different aspects of the same principle that help us understand it from differing perspectives. The first benchmark clarifies *what is meant* by perfection and the second one lays out the process by which we *seek* perfection:

1. Mindset—We challenge our paradigms and expectations. This benchmark focuses our thinking on the ideal itself. What do we mean by perfection?
2. Structure—We approach improvement in a structured way. This benchmark describes the act and process of *seeking* perfection.

The following sections describe each of these behavioral benchmarks in greater detail.

Mindset: We Challenge Our Paradigms and Expectations

Steven Covey said that we should "begin with the end in mind."[10] If the end we are seeking is perfection, how do we define perfection? As

---

[9] Shingo, *Non-Stock Production*, 32.
[10] Steven Covey, *The Seven Habits of Highly Effective People* (New York: Simon & Schuster, 2020), 109.

the Shingo Institute posits, perfection is an aspiration to continuously improve. But is there a universal ideal toward which we can aspire? In the 1990s, in an effort to answer this question, the Toyota Production System Support Center (TSSC) used the term "True North" to define Toyota's vision of the ideal condition for their American suppliers. The term was selected because of its general connotation of an ideal condition and because, like Shingo principles, true north is universal. Regardless of an organization's current condition in its pursuit of enterprise excellence, the "vision of the ideal,"[11] as it was referred to by Hajime Ohba, the General Manager of TSSC, is constant. The TSSC's definition of True North is intentionally prescriptive, focusing equally on the value provided to the customer and on the organizational culture that develops the providers of that value.

Customer Satisfaction

For customers, the absolute conditions of zero defects, zero wastes, and one by one in sequence prescribe, as Ohba noted, "what we *should* do, not what we *can* do."[12] The ideal to produce the customer's exact order immediately and with perfect quality may be, as the Shingo Institute notes, "an aspiration not likely to be achieved," but it is nonetheless *perfection* from the customer's viewpoint. Regardless of industry, these three conditions describe the objectives of continuous improvement. As Shingo notes, they are the end results of consistent application of improvement tools, but they are also the *means* to customer satisfaction.

Human Development

Like the perfect game in baseball, perfect customer satisfaction implies not only the ideal result but also constant pursuit of perfection by every employee to provide that result: "Everyone, every minute, every day," as described by the TSSC model. This means that every employee—executives, managers, and team members—are active thinkers and problem solvers.

---

[11] Hajime Ohba, conversation with Bruce Hamilton, 1996.
[12] Ibid., emphasis added.

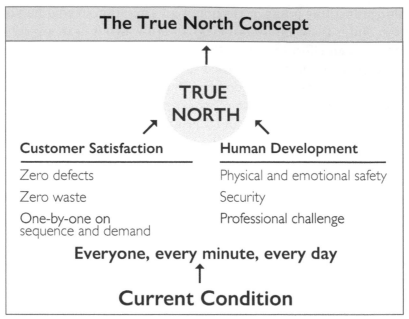

The TSSC True North concept.

Many organizations have tried and failed to create the flywheel of improvement by engaging only a small subset of managers and subject matter experts. Frontline team members receive "awareness" training but are not treated as co-equal thinkers and problem solvers. The outcome is too few problem solvers crushed under the weight of too many problems, while most of the workforce watches from the sidelines. This approach is not sustainable and ultimately leads to disengagement by most team members.

Creating an organizational culture that favors problem identification and develops *every* team member as a thinker and problem solver is *the* challenge that the *Shingo Model* and its fundamental principles address. All ten of the Shingo Guiding Principles support human development by striving to create an environment that is both physically and emotionally safe; where employees feel secure in their jobs; and, in the context of continuous improvement, where all employees are presented with the professional challenge of providing perfect customer satisfaction every minute of every day.

Although they were developed separately, the ideals of the TPS are closely aligned to the Shingo Institute's principles. Organizations outside of Toyota use different jargon to articulate Toyota's philosophy in terms that are more customary to that industry. For example, UMass Memorial Health System's True North philosophy is "best place to receive care, best place to give care." The meaning of this philosophy for patients and care providers captures True North, but the terminology is more fitting to that industry. The words may change, but the fundamental vision of the ideal remains constant. Following the principle to seek perfection as it relates to organizational excellence means embracing both customer satisfaction and human development.

Organizations that deny this principle or subscribe to only parts of it soon plateau in their pursuit of excellence. For example, in November 2008, at the time of the U.S. auto industry bailout, General Manager of General Motors, Rick Wagoner, was asked why they did not subscribe to Toyota's philosophy and methods. He replied, "We're playing our own game—taking advantage of our own unique heritage and strengths."[13] Denying principles has consequences. Remember that Steven Covey said to begin with the end in mind. If that end is perfection, be very clear what you mean by perfect.

In my years as an operating manager and more recently as a consultant, I've observed three types of organizations:

1. Those that are *not* seeking perfection in the True North sense. They, like Mr. Wagoner, are playing their own game and the improvement culture flywheel is not present.
2. Those that are seeking perfection and are headed in the right direction, but they have become complacent after earlier successes and have hit an improvement plateau. These organizations might even be seen as operationally excellent, but they are at risk of losing momentum in the long term. Shingo warns that complacency is the enemy of continuous improvement. We could refer to these organizations as "too happy too soon."
3. Those that are committed to seeking perfection—everyone, every minute, every day—from the frontline to the board room. These organizations are not content to sustain improvement but are seeking to *accelerate* it by having more problem solvers solving more and more problems faster and faster. This is the essence of continuous improvement and this behavior is at the Shingo Prize–award winning level.

---

[13] Alex Taylor, "GM: Death of an American Dream," *Fortune Magazine,* November 2008, https://archive.fortune.com/2008/11/21/magazines/fortune/taylor_generalmotors.fortune/index.htm.

A Pair of Paradigms In his writing, Shingo repeatedly rails against status quo thinking, which exists not only as a set of expectations but also as measurement systems to validate those expectations. In fact, traditional production systems are constructed from multiple paradigms that argue against seeking perfection. Here are two examples that were popular targets for Shingo:

1. **Acceptable Quality Level** (AQL) is defined as the *worst quality level that is still considered tolerable.* The term itself is contradictory as it actually describes the acceptable level of *defects*, and even adjudges the severity of defects—minor, major, and critical—according to their presumed impact on the customer. A level of one half to one percent defects is often baked into quality planning, purchasing standards, and product design, and even embedded in software. AQL automatically excuses a level of quality that is not perfect. And while not stated explicitly, the paradigm assumes that there are diminishing returns in chasing after fewer defects, let alone seeking perfection. Even a Six Sigma standard, which seeks to optimize the quality level (at three parts per million), presumes a level of quality that is less than perfect.

Consider, for example, the common task of purchasing eggs. Most shoppers open the carton to inspect their eggs before buying them. The standard for an average shopper is zero cracked eggs. Even one cracked egg in a million would be unacceptable to the customer, hence the standard practice of inspection. Egg producers and grocers have studied the causes of egg cracks—everything from egg size to the age of the hens to packaging and transportation—and have taken steps to reduce the incidence of cracks, but not to zero. (Incidentally, the AQL for cracked eggs *is* 0, as they are considered to be a health hazard [i.e., a critical defect].) Until the industry achieves zero cracked eggs in every carton, the customer will pay for the cost of cracked eggs as well as the inspection time to assure perfect quality. Shingo introduced prevent-type mistake-proofing and other forms of informative inspection to seek perfection in quality. But the paradigm of AQL persists today in most organizations, legitimizing a level of defects that is not perfect. The consequences to both customer and provider can be immense.

2. **Economic Order Quantity** (EOQ) is another long-standing status quo paradigm that discourages seeking perfection. In this case, it is related to planning and scheduling. Defined *as the optimal number of units that a company should add to inventory with each order to minimize the total costs*

of inventory, EOQ is perhaps the most fundamental paradigm in traditional purchasing and production scheduling. While most organizations still view inventory as a necessary evil to provide good customer service, Shingo viewed it as an *absolute* evil, wastefully produced either too soon or in too large a quantity.

This waste of *overproduction*, Shingo pointed out, was a multiplier of every other waste as every resource needed to be activated unnecessarily. In fact, the EOQ paradigm encourages even more overproduction as manufacturers schedule jobs with similar set-ups together, thereby producing some orders before they are needed while letting orders with dissimilar set-ups wait in queue, even if they are late. The key paradigm shift provided by Shingo was the realization that through pursuit of perfection in machine or process changeover, the customer's exact order can be provided within a fraction of traditional lead times. Traditional manufacturers seek to provide short apparent lead time to the customer through large inventory stores and excess capacity. In hospitals, the excess due to long changeover time is seen in extra patient beds and larger waiting rooms. Shingo's idea was to seek perfection in the improvement of flow and pull, to provide the customer's exact order immediately by making it easy to provide small quantities. When he developed SMED, he challenged a long-standing belief that changeover time, the basis for large lot production, could not be radically reduced. Shingo demonstrated that the application of SMED to machines could actually reduce changeover from hours to minutes; he coined it "a revolution in manufacturing."[14] Yet, in 1989, when Shingo received his honorary doctoral degree from USU, he chided academicians in the audience. "There are 20,000 doctoral dissertations on EOQ," he said, "but not a single one on SMED; this is because SMED is too simple and does not produce enough fodder for PhD theses."[15] Even today, relatively few organizations have embraced the new paradigm entirely, and physical plants and supply chain infrastructure have been designed around an assumption of long set-up times and large production lots.

In the following paragraphs are a few stories of friends and associates of the Shingo Prize that illustrate ideal manager behaviors in fostering new paradigms.

---

[14] Shigeo Shingo, *A Revolution in Manufacturing: The SMED System*, trans. Andrew Dillon (Cambridge, MA: Productivity Press, 1988), title page.

[15] Shigeo Shingo, speech. 14th Partners in Business Operational Excellence Conference, Utah State University, Fall 1988. As recorded by Bruce Hamilton.

CURRENT CONDITION ON THE PAINT LINE

Seeking perfection means treating every current condition as a baseline for improvement. In other words, once we reach a target, that becomes our new current condition as we continue toward the ideal. In 1991, on a visit to the Honda assembly plant in Marysville, Ohio, I learned this lesson from Dave Nelson, an emeritus member of the Shingo Advisory Board and, at that time, VP of Purchasing for Honda. As we passed an automated paint line for car bodies, I noticed that the line changed colors every sixteen parts. "Why sixteen?" I asked. Dave smiled and responded, "Because, presently, if we reduced the lot size to even fifteen, the line would crash. We're just not good enough with our changeovers yet. We know that and we're working on it. Our goal is one by one for color change."

SEEKING PERFECTION AT AN AUTOMOTIVE ENGINE PLANT

Gifford Brown, a former site manager at Ford and emeritus Shingo Advisory Board member, relates the outcome of a set-up reduction exercise at his plant conducted by Shingo in the mid-1980s. While working under Shingo's direction, in just half a day, Ford operators, tool makers, and set-up employees changed over a 300-ton press in under ten minutes, which was down from a previous standard of four hours. When Gifford saw this, he rushed up to congratulate the Ford team.

"Did you imagine this time savings was possible?" he asked one operator.

"Actually, we had already discussed much of what Dr. Shingo showed us," the operator replied.

"Then why didn't you make these changes before now?" Gifford queried.

The answer? "We don't set the standards, boss. You do."

Systems drive behavior. Gifford realized in an instant that seeking perfection is an ideal that must be embraced by all team members, including management. Those closest to the work have the clearest picture of the current condition, but executives and managers have the responsibility to set expectations and manage systems to support team member improvements.

SINGLE PATIENT FLOW

How remarkably different the situation is in organizations that seek perfection. I had the pleasure several years ago of visiting Dr. Sami Bahri at his dental clinic in Jacksonville, Florida. Dr. Bahri, who has been called the World's First Lean Dentist, explained to me how he and his staff pursue the goal of single patient flow.

> We began with the simple procedures, like imaging and cleaning, with the goal of completing the patient's entire procedure in a single visit. Next, we added simple dental procedures, and then crowns. We've adopted quick changeover, pull systems, and other Lean methods to approach the ideal of one-by-one patient flow for every procedure. We're not there yet, but we work toward the ideal every day.

In an industry with no apparent similarities to manufacturing, Dr. Bahri applied the same concepts that Dave Nelson and Gifford Brown did. This time, though, he sought perfect care for his patients.

WHY WE CHALLENGE PARADIGMS AND EXPECTATIONS

So why is seeking perfection from a True North standpoint so important? There are several reasons. First, because the ideal expresses perfection from the *customer's* perspective. A customer who buys a dozen eggs, for example, expects no cracks. Second, because the *goal* of perfection makes necessary the *search* for perfection, which necessitates the use of Lean tools such as set-up reduction, standardized work, poka-yoke, autonomation, and andons. These and many more countermeasures were developed explicitly to give employees the means to reduce waste. Third, as stated in this principle's fundamental truth: "People have an innate desire and ability to improve, which is only limited by their expectations." The pursuit of True North creates a professional challenge for every employee to provide perfect value to the customer—everyone, every minute, every day.

Structure: We Approach Improvement in a Structured Way

Structure is the second behavioral benchmark for Seek Perfection. The search for perfection is not a random search. It is a measured and structured approach. Masaaki Imai describes it as "Kaizen means improvement. . . . Kaizen means *ongoing* improvement involving *everyone*—top management, managers, and workers."[16] A paradox of the TPS is that the *behavior*, the process of seeking perfection, involves many small intermediate changes for the better, each of which is an imperfect solution.

In the words of quality consultant Hiroyuki Hirano, "A 50 percent implementation is fine as long as it's done on the spot."[17]

The ideal behavior noted in a popular Lexus commercial is the relentless *pursuit* of perfection. The Lexus itself may never be perfect, but it will always be the object of the relentless pursuit. We strive for perfection, but we approach it most often in small chunks.

As Benjamin Franklin stated, "Little strokes fell big oaks."

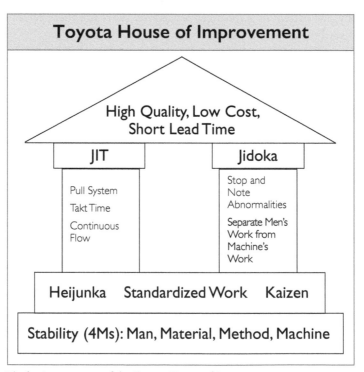

The basic structure of the Toyota House of Improvement.

---

[16] Masaaki Imai, *Kaizen: The Key to Japan's Competitive Success* (New York, NY: McGraw-Hill Education, 1986), xxix.

[17] Hiroyuki Hirano, *JIT Factory Revolution: A Pictorial Guide to Factory Design of the Future* (Portland, OR: Productivity, 1988), 18.

The metaphor of the Toyota House commonly illustrates the necessary order for seeking perfection. At the foundation of the Toyota House is stability. Many organizations begin their enterprise excellence journeys with a great deal of instability, including everything from unsafe operations to unreliable equipment to high employee turnover to design flaws to lack of proper training. These problems create chaos in the work environment that stifles continuous improvement. A team member, for example, who is operating an unsafe machine will necessarily be focused on their own safety rather than on improvement. As noted in Maslow's hierarchy, survival and safety needs will preempt any higher needs. Depending upon the nature of the instability, there are specific countermeasures that may be applied. For example, instability caused by poor organization of work can be remediated by 5S. Or, instability created by machine malfunction or breakdown can be removed through an effective preventative maintenance program.

Above the foundation of stability is the house *floor*, which is comprised of specific tools to create a level flow of clearly defined tasks. Standardized work, for example, is a scientific method for determining the best combination of people, machines, information, and material to fulfill the customer's order. Standardization presumes stability as a prerequisite, because an unstable system cannot be standardized. Similarly, the tools to create flow and quality, as denoted by the pillars of the Toyota house, require that work be standardized first.

Shingo quotes Taiichi Ohno in describing the importance Toyota places on standard work. He writes,

> Standard work sheets and the information contained in them are important elements of the Toyota production system. For a production person to be able to write a standard work sheet that other workers can understand, he or she must be convinced of its importance.[18]

The key point is that seeking perfection should be structured to follow a path that typically begins in chaos and proceeds stepwise to greater target conditions. Continuous improvement tools are essential to organizational excellence, but they are not sufficient to create a culture of sustained improvement. Proceeding in a structured way is critical to seeking perfection.

---

[18] Shingo, *Toyota Production System*, 142.

The following example from my own work demonstrates the importance of structure.

FIRST THINGS FIRST

I experienced this situation many years ago when, as a manufacturing VP, I initiated a set-up reduction project for my company's CNC lathes. According to Shingo, "all set-ups can be reduced by 59/60,"[19] which meant that our set-up times could be ninety seconds rather than ninety minutes. Using Shingo's recommendations from SMED, we succeeded in cutting set-ups down to around forty-five minutes, mostly by getting tools, material, and information ready in advance of the set-up and by cheating (or overriding the production schedule and ganging up like set-ups). After stalling at forty-five minutes for several years, we requested assistance from TSSC to drive the set-up times closer to the Shingo standard. In response to our request, a young industrial engineer visited our facility from a TSSC project company. After interviewing the operators and observing set-ups for a week, she left me a single assignment.

"The BNC [the CNC machine we chose for our set-up project] is not repeatable and needs an overhaul," she said. "I'll be back in three weeks."

I protested, "We can't get it overhauled that fast."

She responded, "Well, good luck then. I guess we're done."

The consultant understood something I had missed: you can't improve an unstable process. I told her we would do our best to overhaul the machine in three weeks and, with some difficulty, we hit the target. On her next visit, operator engagement with the project had increased markedly. At the conclusion of her visit, she left me eight pages of homework.

Within three months of stabilizing our equipment, set-ups were reduced to twenty minutes and we had recalibrated to measuring in seconds. At the nine-month mark, operators were able to change over between any of seventy different parts in less than nine minutes. With further work, the team eventually achieved changeover in under five minutes.

---

[19] Shigeo Shingo, speech. 14th Partners in Business Operational Excellence Conference.

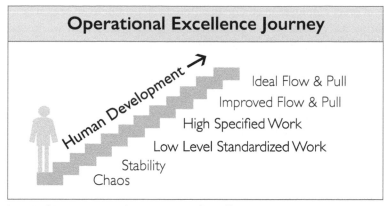

A typical journey toward organizational excellence.

I learned two valuable lessons. First, employee engagement begins with executive engagement. Once I understood what the team members already knew (that the machine was not true), we all became engaged. While our operators may have innate desires, they can be quickly squelched and even turned to spite by disparagement from executives, managers, and supervisors.

Second, I learned that improvement has a natural structure. Whether we choose to accept or deny this aspect of Seek Perfection, there will be consequences. In this case, seeking perfection in changeover began with achieving stability, then standardization, and finally with improvement.

SUPPORTING CONCEPTS

Two key supporting concepts—Integrate Improvement with Work and Identify and Eliminate Waste—provide a deeper perspective on the mindset and structure needed to seek perfection.

The concept, Integrate Improvement with Work, describes the complementary roles of executives, managers, and team members to make improvement *a part of* their jobs rather than *apart from* their jobs. The *Shingo Model* states:

> As the migration toward a principle-based culture occurs, the activities and approaches for continuous improvement become a part of the everyday

work of every employee in an organization. Each person in an organization performs daily work. When improvement is integrated with work, each person naturally accepts responsibility for improvement of the daily work processes.

Executives are responsible for improving strategy-setting processes or perhaps resource-alignment processes. They are primarily responsible for deploying mission-critical strategy and metrics down into the organization so that every person not only has a clear line of sight to what matters most but are also motivated by the mission in a way that creates a compelling case for improvement.

Managers are responsible for improving quality systems, or performance development systems, or value stream flow.

Team members are responsible for improving their cycle times, or quality of work, or yields. They become 'scientists' who continually assess the current state of their processes and pursue a better, future state that will enhance value (or eliminate waste) and, thus, pursue perfection.

Integrating improvement with work is more than assigning responsibility; it entails the creation of standardized work that defines the necessary systems for improvement.[20]

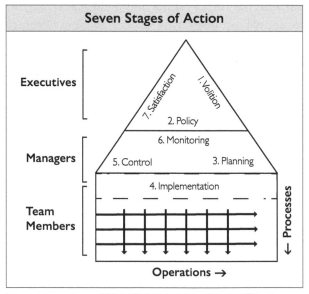

Illustration adapted from Shingo's Seven Stages of Action.

---

[20] Shingo Institute, *The Shingo Model,* 26.

Shingo referred to this division of labor as the "Seven Stages of Action."[21]

1. **Volition.** Top leaders must show a visible commitment to improvement.
2. **Policy.** Clarify the ideal and set direction.
3. **Planning.** Create a plan to reach the ideal.
4. **Implementation.** Carry out the plan.
5. **Control.** Assure that implementation is aligned to the plan.
6. **Monitoring.** Measure and reflect to improve the plan.
7. **Satisfaction.** Evaluate the overall accomplishment and set new goals.

The concept, Identify and Eliminate Waste,[22] is practical for making processes flow, thus it becomes a primary focus of continuous improvement. Waste elimination is a powerful supporting principle because it is easily understood by everyone associated with a value stream (especially in comparison to the complex concepts and computations often associated with cost per unit, cost variances, statistical variability, and other complex metrics). Focusing on the elimination of waste will consistently drive appropriate behavior, while the wrong focus, such as large inventory write-downs, fire sales, or scrap, can frequently become a barrier to improvement.

One way to view waste is to think of it as anything that slows or interrupts the continuous flow of value to customers. In the end, identifying and eliminating waste is a concept that effectively engages the entire organization in the continuous improvement effort.

The seven wastes identified by Shingo provide a tangible differentiator between work and all the things that get in the way of work. For example, storage, transport, overproduction, processing, motion, defects, and waiting are all industrial engineering wastes that typically represent more than 95 percent of the elapsed time to provide a good or service, but they add no value. These wastes, however, are often hiding in plain sight; team members don't see them because they are so embedded in the job.

"The most damaging kind of waste is the waste we don't recognize," Shingo noted.[23]

The search for perfection requires that team members and management actually *see* the waste. Any reasonable person will remove waste if they can see it. The problem is, they often don't. The title of the book, *Learning to*

---

[21] Shingo, *Non-Stock Production*, 211.

[22] Shingo Institute, *The Shingo Model*, 25.

[23] Shingo, *The Production Management System*, 35.

See,[24] by John Shook and Mike Rother, alludes to this challenge. A worker who is searching for a wrench to set up a machine or a doctor running up a staircase to see a patient, for example, may each feel like they are working, but in actuality they are experiencing motion waste.

Pursuit of perfection requires continuous surfacing and removal of waste by "everyone, every minute, every day." The fourth stage of Shingo's Seven Stages of Action, implementation, is designated as a frontline responsibility (see the figure on page 42). Critically, the fourth stage will not gain traction without collective efforts in the other six stages of action.

Reader Challenge

Consider the work you do and try to articulate what perfection would look like in your work. Who are the recipients of the ideal outcome? List the key work systems to provide the perfect results. Who are the providers? How would you rate the current outcomes versus your image of the ideal? What are the obstacles to achieving perfection?

Reflect on the use of the word *optimal* as it describes results for key systems in your work. What does optimal mean? Does it create a conflict with seeking perfection?

SUMMARY

Without an ideal condition to pursue, improvement will be finite, which is the unfortunate case for many organizations. Perhaps the most common complaint by Lean implementers is the difficulty in sustaining gains. Seeking perfection means committing to an ideal: what we *should* do, not just what we *can* do. Seeking perfection is also creating a structured approach in which executives, managers, and team members each play a necessary role. The principle creates a mindset of never-ending improvement that unlocks the "innate desire and capability" of every team member.

---

[24] John Shook and Mike Rother, *Learning to See: Value Stream Mapping to Add Value and Eliminate Muda* (Boston, MA: LEI, 1999), cover.

4

Embrace Scientific Thinking

*Any improvement must be made in accordance with the scientific method,
under the guidance of a teacher, at the lowest level possible in the organization.*[1]

—**Steven Spear and H. Kent Bowen**

BUSINESS CASE FOR EMBRACE SCIENTIFIC THINKING

Innovation and improvement are the consequence of repeated cycles of experimentation, direct observation, and learning. A relentless and systematic exploration of new ideas, including failures, enables one to constantly refine their understanding of reality.[2]

WHAT IS A FUNDAMENTAL TRUTH?

Undergirding every principle is a fundamental truth that provides the basis for why a principle is universal and timeless. The truth becomes self-evident once it is discovered.

A fundamental truth held by the Shingo Institute is the following: *The best decisions are based on a clear understanding of reality.*

---

[1] Steve Spear and H. Kent Bowen, "Decoding the DNA of the Toyota Production System," *Harvard Business Review*, September–October 1999, 98.

[2] Shingo Institute, *The Shingo Model*, 22.

DOI: 10.4324/9781003216698-5

A fundamental truth deepens our understanding of a principle. It becomes part of our values and beliefs, and it influences our behavior. Therefore, it should not be taken lightly. The Shingo Institute advocates for more than one fundamental truth in an organization. In fact, there can be many. Consider another fundamental truth: all employees can solve problems.

What might we expect to see in an organization that shared this fundamental truth?

Perhaps we would see executives seeking out and recognizing manager and team member behavior in support of the principle Embrace Scientific Thinking. We would likely see managers allocate time in their schedule to develop and improve systems that support the principle Embrace Scientific Thinking. We may also see managers encouraging team members to be involved. And we would likely see team members behaving as active participants.

Reader Challenge

Develop a personal fundamental truth for the principle Embrace Scientific Thinking. Consider using a *because/therefore* statement. The fundamental truth is the *because* part of the statement. The principle is the *therefore* part of the statement. Here's an example: Because all employees can contribute to the success of the organization (*a fundamental truth*), therefore, I respect every individual (*principle*).

Because _____ (*my fundamental truth*), therefore, I embrace scientific thinking (*principle*).

THE SUPPORTING CONCEPTS

Four supporting concepts, Stabilize Processes, Go & Observe, Standard Work, and Rely on Data and Facts, provide an in-depth perspective on the mindset and behaviors needed to Embrace Scientific Thinking.

Stabilize Processes: *Stability in processes is the bedrock foundation of any improvement system. It creates consistency and repeatability and is the basis for problem identification. Nearly all continuous improvement principles rely on stability because it is the precursor to achieving flow. Many of* *the rationalizations for waste are based on the instability of processes, as if they are beyond our control. Instead, organizations should apply the basic tools available to reduce or eliminate instability and thus create processes that help to identify and eliminate waste.*[3]

The foundation of Lean is a stable and standardized process. It represents the important early work in moving from a chaotic environment to a JIT environment. Creating a more stable and standardized environment is a prerequisite for JIT. It can't be skipped. Will a kanban replenishment system work in an environment with quality problems and material issues? Will set-up reduction work in an environment with equipment and staffing issues? Stability is typically found in the 4 Ms (man, method, material, and machine). A level of stability is required in each of the 4 Ms in order to do the work. People need to be trained and arrive on time each day. Equipment needs to be available and in good working order. A steady stream of good quality material is needed, and the method needs to be known and followed. In the context of the principle Embrace Scientific Thinking, organizations have two paths when they are faced with stability issues. They can firefight and work around problems that continually reoccur, or they can apply scientific thinking to investigate the cause of problems and implement countermeasures to prevent them from recurring.

Remember that Shingo quotes Taiichi Ohno in describing the importance Toyota places on standard work. He writes,

> Standard work sheets and the information contained in them are important elements of the Toyota production system. For a production person to be able to write a standard work sheet that other workers can understand, he or she must be convinced of its importance.[4]

There cannot be true improvement without standards. Trying to improve an environment with lots of variation is like trying to improve a moving target. Every time you look at a process it is different, and every person you observe has a different way of doing things. Trying to improve in an environment with lots of variation is nearly impossible. In terms of processes, only one method of doing the work should be the agreed-upon method. You can't have multiple ways of doing things and expect to get the same result. Scientific thinking and scientific methods are used to eliminate variation and to develop the best-known method for work.

---

[3] Ibid., 23.
[4] Shingo, *Toyota Production System*, 142.

Go & Observe: *Direct observation is a supporting principle tied to scientific thinking. It is, in fact, the first step of the scientific method. Observation is necessary to truly understand the process or phenomenon being studied. All too frequently, perceptions, past experience, instincts, and inaccurate standards are misconstrued as reality. Through direct observation, reality can be seen, confirmed, and established.*[5]

We should always insist on direct observation. The power of this concept is obvious every time it is practiced. Observation provides first-hand knowledge, a basis for reality, and a motivation unparalleled by other improvement practices. Organizations that practice this concept will exhibit behaviors such as executives systematically visiting all work areas, managers being present as a matter of practice in the work area, and team members being provided the necessary tools to record their continuous observations. Of course, managers and executives will not be able to observe all occurrences of issues; only those who do the work can comprehensively understand what is going on. Having said that, managers and executives need to be sufficiently present to actually see the process, and the process has to be obvious enough for visibility to occur quickly.

Standard Work and Standardized Work: *While stability is a necessary precondition for creating flow and improvement, creating standard work builds control into the process itself. Standard work is the supporting principle behind maintaining improvement rather than springing back to preceding practices and results. Standard work also eliminates the need to control operations through cost standards, production targets, or other traditional supervisory methods. When standard work is in place, the work itself serves as the management control mechanism. Supervisors are more free to work on other tasks when there is no need to monitor and control the work process.*[6]

SOPs document the agreed-upon method. They are a work instruction that defines how to do something. Examples include how to generate a quote, how to enter an order, how to check in a patient, how to set up a press, how to answer a customer call, how to perform a test, and many others. Team members who do the work should be involved in determining the standard or the agreed-upon method. The aim is to eliminate variation while developing the best-known method by engaging the people who do the work.

When team members are involved in the development of standard work, it becomes *their* standard. If they are not involved, it becomes

---

[5] Shingo Institute, *The Shingo Model*, 24.
[6] Ibid., 24.

someone else's standard. Scientific Thinking is practiced through direct observation, measurement, discussion, experimentation, adjustment, and re-measurement. Determining the agreed-upon method should be a deliberate and thoughtful activity. But remember that it is not a forever standard; there should always be an expectation that the work will improve.

Standardized work, on the other hand, is more than an SOP or work instruction. It is a work plan that better matches *how we work* instead of a work instruction that tells us *how to do* something. Standardized work is the best combination of resources to meet customer demand with a quality part or service.

A simple example might help illustrate the meaning and purpose of standardized work. If you were to enter a fast-food sandwich restaurant at 10 A.M. (an off-peak time), there may be only one person working the counter. They will take your order, make your sandwich, place it in the oven, add condiments, package your food, and cash you out. Why is only one person working the counter? Because the frequency, or pace, of customers entering the store compared to the cycle time of making a sandwich dictates that one person should be able to handle all the responsibilities. At 11 A.M., there may be two people needed to complete the process. One person to take the order, make the sandwich, and place it in the oven; and a second person to remove it from the oven, add condiments, package the food, and cash you out. At 11:30 A.M., there may be three people working the process. Finally, at peak lunchtime, there may be four people working the process.

Standardized work develops and documents the best work sequence for different levels of customer demand—in this case, when the restaurant is busy and when it is slow. It defines and documents the best combination of resources to meet customer demand. Scientific thinking is embedded in the process of developing standardized work. Therefore, standardized work is a scientific method.

Rely on Data and Facts: *Dr. Shingo emphasized the importance of being data-driven in the pursuit of continuous improvement. He frequently shared examples of specific situations where data was collected, but the data was incorrect or wasn't actually used in the improvement process. He was adamant that the understanding of the actual process be so thorough that when implementing a change in the process, the improvement, as evidenced by the data, could be predicted. Thus, reconciliation would be required between the predicted results and the actual results, making the improvement process truly data driven. Ultimately, when data is treated loosely or imprecisely,* *the tendency is to leave potential improvement on the table or, even worse, to not achieve any improvement at all.*[7]

One definition of the word *fact* is "something that has actual existence."[8] The supporting concept, Rely on Data and Facts, requires all members in an organization to see, feel, touch, and hear the reality of the process. Shingo says that if you have not seen it for yourself, it is not a fact.

The primary behavioral sign of this concept in action is the way people talk about opportunities and issues. Without data, people will often use terms such as "I think" or "they said." In contrast, *data* is defined as "factual information (as measurements or statistics) used as a basis for reasoning, discussion, or calculation."[9] Data is the representation of the fact in a way that enables others to see and understand it. Executives and managers must provide team members with simple, visual mechanisms to present facts that are associated with their part in the overall process.

Problem Statements Without Data	Problem Statements With Data
Rusty converter covers	30–40% of covers received from one supplier are rusty. (0% received from other suppliers.)
Burrs created by gear shaping	30–50% of burrs created by gear shaping are not removed.
Coils burning up	5 machine coils burned up in the last month, leading to 20 hours of down-time.
Incomplete / inaccurate information on customer returns	40% of customer return forms have incomplete and/or inaccurate information.

FIGURE 4.1
Data adds depth and credibility to the problem statements.

---

[7] Ibid., 27–28.
[8] *Merriam-Webster's Collegiate Dictionary,* 11th ed., s.v. "fact."
[9] Ibid., "data."

Data and facts take problems from being perceived as subjective, or an opinion, to being objective, or real. Consider the set of problem statements in Figure 4.1. Four problem-solving teams at a manufacturing company I was working with developed the statements on the left side of the table. After a quick discussion with each team, I urged them to develop a second set of problem statements, which are shown on the right side of the table. Did data add credibility to the problem statements? Did data move the problems from being subjective to objective? Which statements do you think would gain more support? What would be the likely response for each set of statements?

BEHAVIORAL BENCHMARKS AND RELATED CONTEXT

A deeper understanding of the Embrace Scientific Thinking principle is provided through the following behavioral benchmarks:

1. **Reflect:** We understand that decisions and changes are based on careful examination of challenges and opportunities.
2. **Analyze**: We experiment, innovate, and make decisions with an appropriate analysis of good data and facts.
3. **Collaborate**: We actively seek insight and ideas, especially from those closest to the work.

The four continuous steps PDCA (plan, do, check, act) were made popular by W. Edwards Deming, the father of quality improvement. After many years, Deming modified PDCA to PDSA (plan, do, *study*, act), because he believed the *check* phase emphasized inspection over analysis. *Study* more accurately defined the stage after *doing*, meaning that we must study the results of the *do* phase. Reflection is a part of the study phase and is a natural transition from one experiment to another or from one improvement to another. Did the experiment provide the expected results? Did it produce an unanticipated benefit or cause an unanticipated burden? What worked? What didn't work? Reflecting on these and other questions leads to additional experiments or improvements, extends the PDSA cycle, and embodies the word *continuous* in continuous improvement.

The Merriam-Webster's Collegiate Dictionary lists nine definitions for the word *reflection*. Definition number two states that reflection is "the production of an image by or as if by a mirror."[10] Applying this meaning provides a level of personal responsibility as it pertains to problem solving and improvement work. Let's take a brief look in the behavior mirror to see our own personal support for decisions and changes. I may ask myself: Have I done enough to support change and improvement? Do I seek input from others? What could I do to improve the problem-solving process, the improvement process, and the decision-making process? The process of personal reflection also connects directly to the principle Lead with Humility.

The dictionary's sixth definition states that *reflection* is "a thought, idea, or opinion formed, or remark made as a result of meditation." This is the essence of reflection as it pertains to the scientific method. Each step of the PDSA cycle requires time, effort, and reflection. The steps should not be taken lightly. How much effort does an executive, manager, or team member exert in planning, doing, studying, and acting on changes? Shingo Prize examiners look for ideal behavior, but they often find less than ideal behavior as it pertains to supporting and/or contributing to changes and improvements.

The 1999 *Harvard Business Review* case, "Decoding the DNA of the Toyota Production System," by Steven Spear and H. Kent Bowen, provides insight into the unspoken rules that gave Toyota its competitive edge. The rules became known as the Four Rules. Rule 4 states, "Any improvement must be made in accordance with the scientific method, under the guidance of a teacher, at the lowest level possible in the organization." The article explains,

> Few organizations have managed to imitate Toyota successfully—even though the company has been extremely open about its practices. We found that, for outsiders, the key is to understand that the Toyota Production System creates a community of scientists . . . following the scientific method.[11]

Reader Challenge

Re-read the supporting concept Go & Observe, which was discussed earlier in this chapter. Consider the word *reflection* and its second (looking in a mirror) and sixth (a thought, idea, or opinion) definitions

---

[10] Ibid., "reflection."
[11] Spear and Bowen, "Decoding the DNA of the Toyota Production System," 98.

as they pertain to insisting on direct observation. Consider the maturity lenses used by examiners when they are assessing behavior. How would you answer the following questions?

Frequency: How often do I practice the behavior?

Duration: How long have I been practicing the behavior? Is this the first time or have I practiced it for many years?

Intensity: Do I have a passion for this behavior? Do I place importance in the behavior?

Scope: Are many others involved in this behavior or is it limited to just a few individuals within the organization?

Role: Are executives, managers, and team members equally involved?

The difference between the observed behavior (as seen by an examiner) and the ideal behavior (as known by an examiner) is called the *behavioral gap*. What behavioral gaps have you identified after reflecting on the supporting concept Go & Observe? List your observations below:

_____

_____

_____

_____

_____

Many years ago, I took my then 5-year-old daughter to a local science museum. It was a great day. She was naturally curious, inquisitive, and open minded. She seemed to love science and learning. She's now 26 years old and her interest in science continues. For most children, scientific thinking comes naturally. It feeds our intuition and desire to learn and grow. We observe, ask questions, experiment, analyze and interpret data, develop explanations, engage in discussion, and design solutions. In other words, this innate behavior in children represents Scientific Thinking. While children may not know what any of those terms mean, they naturally apply them—and so can adults!

Applying scientific thinking to how people work, how people improve work, and how people solve problems is at the heart of the principle Embrace Scientific Thinking. Thinking is intuitive for all adults. It's a built-in skill. It comes naturally. In fact, adults solve hundreds of problems every day. What time to wake up, what to wear, what to have for breakfast, whether to bring a lunch or eat out, whether to bring an umbrella along, which roads to travel, where to park. The list goes on. Yet many organizations discourage problem solving by most team members, either explicitly or implicitly. They tend to have thinkers and doers. A small percentage of team members are asked to solve problems, while everyone else is told to "put in a good day's work" and "leave the problems for others to solve." Most organizations have way too many problems and way too few problem solvers. Traditional organizations pick and choose their problem solvers. Some are authorized, while others are not. Whole departments may be excluded from the problem-solving process. Does this sound familiar?

Developing people is at the heart of the *Shingo Model*. Team members are not solving problems for the sake of it. They are not solving problems because they took a Problem-Solving course or read a book. They are solving real problems, company problems, customer problems, supplier problems, and process problems. They are solving problems they encounter while trying to get their job done. The *Shingo Model* brings problems to the surface. It exposes problems that may or may not be understood by everyone. This creates the need for problem solving. More and more problems being exposed requires more and more problem solvers. The corollary is that team members respond to the need and engage in problem-solving activities. All team members are expected to solve problems quickly and effectively. They are also expected to use scientific thinking and a scientific method. See Figure 4.2.

Applying scientific thinking to how we work, how we improve our work, and how we solve problems is accomplished through the scientific method.

Create the Need →	Respond to the Need
Bring Problems to the Surface	**Engage in Problem Solving**
Equipment issues	Solve problems quickly and effectively
Long set-ups	
Rework	Develop team members on the shop floor
New hires	Embrace scientific thinking
Unclear instructions	Everyone, every minute, every day
Missing parts	Your competitive advantage
Mistakes	

FIGURE 4.2
Developing people.

It is not a random, unstructured process. There is a structured method for it that could be intuitive or instinctive for some problems but must be explicit for others. Scientific thinking is involved in both scenarios. A simple problem might be solved with intuition. For example, you notice that a light is out in your kitchen. Instinctively, you develop a hypothesis and begin testing it. Are there other lights out in the house? Could it be a burned-out bulb, or could it be caused by something else? After determining that it is the only light out in the house, you might try a test by changing the bulb to see if it fixes the problem. If it works, you've solved the problem scientifically, albeit through intuition.

But not all problems can be solved through intuition or instinct. Some problems and improvements require a more formalized use of the scientific method. They require an explicit use of assumptions, hypothesis, experimentation, analysis, and adjustment. For example, a key part for an assembly has been out of stock several times in the past six months. You could repeat the countermeasure that was developed via intuition or instinct or you could try a more explicit use of the scientific method to uncover the real problem, or the root cause, and develop a countermeasure to prevent it from re-occurring. This method might look something like Figure 4.3.

The 5 Whys of Lean Problem Solving

1. Why was the order late?
   → It was missing a part.

2. Why was the part missing?
   → Machining didn't complete it on time.

3. Why didn't machining complete it on time?
   → The machine broke down.

4. Why did the machine break down?
   → It was leaking oil.

5. Why was the machine leaking oil?
   → It was not on a PM program.

Countermeasure: Put the machine on a PM program to prevent breakdowns.

FIGURE 4.3
An example of the 5 Whys of Lean problem solving.

The 5 Whys is a scientific method for Lean problem solving and for finding the root cause of a problem. Can you think of other scientific methods for problem solving or for structured improvement? Would you consider A3, 8D, DMAIC, CEDAC, standardized work, and improvement kata to be scientific methods? I would. Scientific methods provide the means for structured problem solving and for improvement. Without them, we would struggle to fix anything. Without them we would struggle to make meaningful and sustainable change.

Remember that not all problems require the rigor of a formal scientific method. But other problems do. Some organizations try to solve problems and make improvements without the guidance of a scientific method. Then they wonder why problems are not solved and improvements are not sustainable. Workplaces are complicated. Processes are complicated. Therefore, they require an explicit and formal use of scientific thinking and scientific methods.

Reader Challenge

Consider the second and third Insights of Organizational Excellence: *Purpose and Systems Drive Behavior* and *Principles Inform Ideal Behavior*. Think about what these statements mean as they relate to the principle Embrace Scientific Thinking.

1. Try to articulate what "Principles Inform Ideal Behavior" means as it relates to the principle Embrace Scientific Thinking. Think about ideal behaviors of executives, managers, and team members as it relates to this principle.
2. Try to articulate what "Purpose and Systems Drive Behavior" means as it relates to the principle Embrace Scientific Thinking.
3. Now consider your current situation for problem solving. What principle are you living by as it relates to problem solving? Are there systems for problem solving? Are they informal or formal? What behavior is the system driving? What behavior can be observed from executives, managers, and team members as it relates to problem solving?

CHIEF PROBLEM SOLVER

Bruce provides the following anecdote:

Earlier in my career, I was often asked to be part of a project team or problem-solving activity primarily due to my title. As an engineer, I was deemed a problem solver, which is typical in most workplaces. As a supervisor and manager, I became the Chief Problem Solver for my group. Problems were passed my way and would pile up on my desk waiting for attention. I worked to solve them between production meetings, project meetings, quality meetings, planning meetings, and other meetings. I triaged them to work on the most important problems. Problems that were holding up orders and shipments needed my immediate attention, while other problems could wait. Problems I couldn't get to today would be there tomorrow. Predictably, problems piled up faster than I could solve them. I became a bottleneck. I had to do something fast. I could ignore them, I could cut corners, or I could share them. If I shared them, I would increase problem-solving capacity by enlisting others to be problem solvers.

A good first step is to go to the workplace and ask team members for help, which is an example of Leading with Humility. Asking questions, soliciting help, and accepting offers for assistance was the only strategy that made sense to me.

Engaging others in problem solving worked for some individuals but not others. A few people were eager to get involved, while others were hesitant.

After some reflection, it has become clear to me that a verbal request for support is not enough for most people to change their behavior. It was treated as simply a passive request that did not provide a structure or system. Systems provide the means for everyone to become problem solvers. They change the request for involvement from passive to active. Systems guide the process and ensure a better outcome. Management's role is to develop systems to surface and solve problems quickly and effectively, and to involve everyone. This is the competitive advantage—developing problem-solving capability and capacity.

Reader Challenge

Review Figure 4.4 and determine a problem-solving level for most team members in your department or organization. You might consider polling coworkers to get their input. What conclusions can you

Problem-Solving Maturity Table	
Level	**Description**
5	All team members actively engaged in problem identification and solving
4	Reported after countermeasure implemented
3	Problems reported and solutions offered
2	Problems reported
1	Problems not reported
0	Problems hidden due to fear

FIGURE 4.4
Problem-solving maturity levels.

draw from the responses about the state of problem solving in your organization? Is there room for improvement?

Level Description

0. At the lowest level, problems are hidden due to fear.
1. Problems are not reported due to apathy. Management doesn't seem to care, so why should I?
2. Problems are reported, but solutions are not offered even if they are known. There's a clear division of labor where managers solve problems.
3. Team members are more engaged. They report problems and offer solutions.
4. Team members are encouraged to solve problems and the organization has systems in place for it.
5. All team members are actively involved in both problem identification and problem solving, and there are systems in place for it.

Enter the current level of problem solving for most team members in your department or organization: _____.

Return to this activity in one year and reconsider the level of problem solving for most team members in your department or organization. Enter your new level here: _____.

Did the level stay the same or did it improve? If the level stayed the same, why? If it improved, how? Finally, what could you do to improve your level? Think of the principle and insights as you consider your answer.

KEY SYSTEMS TO SUPPORT EMBRACE SCIENTIFIC THINKING

A few key systems that closely align with the principle Embrace Scientific Thinking are listed in the following.

1. **Coaching and Mentoring System:** Executives and managers must make a personal commitment to developing others through coaching teams and mentoring individuals. Tools, such as A3 Thinking and Improvement Kata, provide the means for it as a method for communication.

2. **Kaizen (Continuous Improvement) System:** Executives and managers must set an expectation that the status quo is not acceptable, and improvement is a normal part of how we work. Individuals and teams should make improvement in their own areas first. This is called *process kaizen*. As experience is gained and broader improvement is needed, individuals and cross-functional teams can make improvement in up and down stream areas. This is called *system kaizen*. The kaizen system provides a standard framework for improving both work systems and management systems, which are the primary systems for operating and overseeing any organization. A wide range of improvement tools can be used in both process kaizen and system kaizen. Tools provide the means to improve processes and systems. Without them, individuals and teams will struggle.

3. **Problem-Solving System:** Problems must be solved through a structured and scientific method that embeds planning, experimenting, checking, and acting or adjusting into the culture. Managers must develop tools that encourage team member participation in the problem-solving effort.

4. **Quality Improvement System:** Organizations need to move away from traditional quality control activities (e.g., inspecting, sorting, and screening activities to remove defects) and develop quality

systems that utilize improvement tools to build quality into the process. Tools typically found under the principle Assure Quality at the Source should come to mind. Each of these is a scientific method with embedded scientific thinking.

This list is not meant to be comprehensive. It provides a few of the more obvious systems to consider as they relate to Embrace Scientific Thinking. Many other systems within your organization certainly merit close scrutiny as you embrace scientific thinking throughout the organization.

TOOLS TO SUPPORT EMBRACE SCIENTIFIC THINKING

A few key tools that closely align with the principle Embrace Scientific Thinking are listed in the following.

1. **A3 Problem Solving** is a structured problem-solving and continuous-improvement approach, first employed at Toyota. It provides a simple and standardized approach to problem solving using a single sheet of A3-size paper (approximately 11 × 17).
2. **5 Whys** is the most basic form of root-cause analysis. It involves asking *why* five times to get to the root of a problem.
3. **6-Step Problem-Solving Method** is a basic model for defining and solving problems: 1) define the problem, 2) analyze, 3) brainstorm potential solutions, 4) select a solution and create a plan for implementation, 5) implement the solution, and 6) evaluate the solution.
4. **8D Problem Solving** is a problem-solving method used by the U.S. government for its military operations and was later adapted and popularized by Ford Motors. As the name implies, 8D has eight disciplines that form the basis for an 8D report: 1D) team formation, 2D) problem description, 3D) containment actions, 4D) root-cause analysis, 5D) corrective actions, 6D) validate corrective actions, 7D) identify and implement preventive actions, and 8D) team and individual recognition.
5. **CEDAC** is a team-based problem-solving method based on the fishbone diagram. It is enhanced through a systematic evaluation process using color-coded cards. CEDAC stands for **Cause-and-Effect Diagram with the Addition of Cards.**

6. **DMAIC** is a data-driven improvement method used to improve quality: Define, Measure, Analyze, Improve, and Control. DMAIC is typically associated with a Six Sigma approach to improvement.
7. **Improvement Kata** is a scientific method for breaking challenging goals into daily actions. It is a popular method for coaches and learners.
8. **Standardized Work** is one of the key components of JIT. This method carefully develops a work sequence and standard WIP (work-in-process inventory) to match the expected rate of customer demand (takt time). A focus on motion and machine improvement is critical to providing the greatest productivity and the highest quality while making the job easier and safer.

This list is not meant to be all encompassing. It provides a few of the more obvious tools to consider as they relate to Embrace Scientific Thinking. There are certainly many other tools that may merit close scrutiny as you embrace scientific thinking throughout the organization.

SUMMARY

A deep understanding of the principle, Embrace Scientific Thinking, is essential if you plan to achieve the full benefits of the *Shingo Model* within your organization. This principle reflects the basic thinking that we must continuously develop our organization and our people to effectively and efficiently improve for the long term. Problem solving and improvement must move beyond the people at the top. All team members are willing and able to contribute. Systems and tools move the aim from a passive request for involvement to an active request.

Remember that people run the system, and the system runs the business. However, without equal and appropriate attention to both the social and the technical side of improvement, success will be limited.

The more deeply we understand the *Shingo Model*, its Guiding Principles, and the Three Insights of Organizational Excellence, the more we drive ideal behavior. And the more we drive ideal behavior, the closer we are to achieving long-term, sustainable success.

5

Focus on Process

The conventional approach to production improvement has been rooted in the mistaken notion that production is equivalent to operations. People have not realized that production is a network of processes and operations.[1]

—**Shigeo Shingo**

BUSINESS CASE FOR FOCUS ON PROCESS

The *Shingo Model* explains the principle Focus on Process: *All outcomes are the consequence of a process. It is nearly impossible for even good people to consistently produce ideal results with poor processes. It is human nature to blame the people involved when something goes wrong or when the resulting product or service is less than ideal. But in reality, an issue is usually rooted in an imperfect process, not in the people involved.*[2]

FUNDAMENTAL TRUTH

Great processes set people up to succeed.

---

[1] Shingo, *The Shingo Production Management System*, 12.
[2] Shingo Institute, *The Shingo Model*, 22.

DOI: 10.4324/9781003216698-6

DEVELOP GREAT PROCESSES

Delivering great processes to team members is the responsibility of executives and managers. Executives should concentrate on developing and sustaining a culture that will result in great processes. The Shingo Institute suggests that this should consume 80 percent of executive time, effort, and attention. Managers have an obligation to develop and provide team members with processes and systems that set people up to succeed. Again, the Shingo Institute indicates that 80 percent of manager time, effort, and attention should focus go to this endeavor.

The behavioral focus of this principle is on developing processes. Once accomplished, it will eliminate the tendency to blame people when problems occur. If an organization is to achieve ideal results, eliminating blame is a fundamental requirement of executives and managers.

DON'T BLAME OTHERS

A common practice when something goes wrong is to blame others. Blame begets blame. Eventually, it can result in blaming people in all directions. Management can blame team members when issues arise. Likewise, team members can blame "the system," and thus the managers who developed it, when issues arise. Why, then, is there a natural tendency to blame other people when something goes wrong or is less than ideal?

Dr. W. Edwards Deming introduced leaders around the world to the problems with this type of thinking with his red bead experiment.[3] In the red bead experiment, there is a fundamental flaw in the process that introduces red beads (defects) among white beads (good products). Leaders try various different management practices to drive improvement. First, they try incentives. But the incentives don't work. Then they try recognition. But recognition doesn't work. Eventually, they try threats (or blame). Of course, threats don't work either. None of these approaches address the fundamental problem of a flawed process.

Research indicates that the vast majority of problems are caused by a flaw in the process. Basically, most people want to do a good job—do the right thing—but the process sometimes makes it hard to do the right thing and easy to do the wrong thing.

---

[3] W. Edwards Deming, *The Red Bead Experiment with Dr. W. Edwards Deming,* The Deming Institute video, 9:00, https://deming.org/deming-red-bead-experiment/.

Blaming others not only doesn't solve the fundamental cause, but also leads to other bad behaviors. People will hide problems in order to avoid being blamed. As a result, the problems start compounding. People start hoarding information because they are afraid it will be used against them. As a result, everyone else understands less and less about what the problems are. Eventually, as is a frequent outcome of the red bead experiment, people start cheating to avoid being blamed and potentially being fired.

Leaders need to focus the organization away from these wasteful negative behaviors and focus instead on improving the process.

WHO IS RESPONSIBLE?

Who is responsible for the failing process? The Shingo Institute teaches that executives and managers are responsible for processes, and that it is their responsibility to make sure the processes are designed so that it is easy to do the right thing and hard to do the wrong thing.

When using the *Shingo Model* as a framework, organizations strive to achieve a state where observed behaviors create and sustain processes that result in success for all team members at all levels of the organization. If an organization is looking for sustainable, excellent results, the focus will require a consistent focus on behaviors.

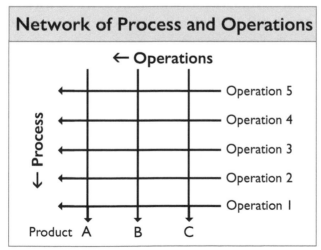

In Shingo's network of process and operations, process is the stages through which material or information gradually move to become a completed product or service. An operation is the discrete stage at which a worker may perform some portion of the work.

PROCESSES VERSUS OPERATIONS

How do we do this? Shingo's perspective on process introduces the technical knowledge required of us. He said:

> Production is a network of operations and processes, with one or more operations corresponding to each step in the process.[4]
>
> To make fundamental improvement in the production process, we must distinguish product flow (process) from workflow (operations) and analyze them separately.[5]

Shingo goes on to further describe the distinction between processes and operations:

> A *process* is a flow by which raw materials are transformed into finished products. Processes play the primary function in production—what we have called the 'frontal view' function. Next are operations, functions we need to regard as secondary and complementary to process functions.
>
> Process functions can be expressed by four different phenomena: 1) processing, 2) inspection, 3) transportation, and 4) delays.
>
> *Process*: A course of events by which materials are changed into products. *Operation*: Work performed by people and machines on materials. Before the division of labor came along, operation and process were embodied in the same worker. The distinct phenomenon of process was not recognized. However, with the division of labor, the flow of operations became clearly separated from the flow of processes.[6]

PROCESS FIRST, THEN OPERATIONS

Once processes are separated from operations, Shingo directs that improvement efforts should focus on the process first. Only after process improvements are complete should improvement efforts then focus on operations.

---

[4] Shingo, *A Revolution in Manufacturing*, 5.
[5] Shingo, *A Study of the Toyota Production System*, 4.
[6] Shingo, *Non-Stock Production*, 232.

First *process* and then *operations*. While processes flow through the plant, however, an operation is performed in one spot and involves actually shaping the product. This is why we are more aware of operations and why process functions and phenomena end up escaping our attention. The result has been the delusion that production is synonymous with operations.[7]

The conventional approach to production improvement has been rooted in the mistaken notion that production is equivalent to *operations*.[8]

The realization that production is a network of processes and operations frees us of our obsession with streamlining *operations* and focuses our attention on making *process* more rational. The result is unprecedented improvement.[9]

Reader Challenge

Go to where the work is done and document the process you observe. Seek a deep understanding of the process. Now identify the time committed to each of the process elements and determine the waste in the process. Continue the investigation by extending the value stream into sales, engineering, and support areas to fully understand the impact on customers.

THE SUPPORTING CONCEPTS

The two key supporting concepts, Go & Observe and Stabilize Processes, provide a deeper perspective on the mindset and behaviors we need to focus on process.

Go & Observe: *Direct observation is a supporting principle tied to scientific thinking. It is, in fact, the first step of the scientific method. Observation is necessary to truly understand the process or phenomenon being studied. All too frequently, perceptions, past experiences, instincts, and inaccurate standards are misconstrued as reality. Through direct observation, reality can be seen, confirmed, and established.*[10]

---

[7] Shingo, *The Shingo Production Management System*, 55. Emphasis added.
[8] Ibid., 12. Emphasis added.
[9] Ibid., 3. Emphasis added.
[10] Shingo Institute. *The Shingo Model*, 24.

Go & Observe requires us to be where the work is done, and to primarily watch the flow of value.

It is best to look at Go & Observe as an "undertaking," not an activity. Periodic visits to where the work is done, even with great frequency, will be of minimal benefit unless work has been realized to incorporate *all* the supporting concepts and principles. For instance, periodic visits to the workplace in an unstable environment will diminish the quality of the observation. Likewise, a workplace without clear visual indicators will be less observable. Consider the dictionary's definition of *observe*: "to watch carefully especially with attention to details or behavior for the purpose of arriving at a judgment."[11]

Ritsuo Shingo taught in a Go & Observe discussion:

> You should have a very big eye—I mean you should watch carefully. And you should have a very big ear. That means you should listen very carefully. But not a big mouth. The leader should be a very good listener. If you talk too much to the people, they just listen to you and try to do what you say. No! The opinion from the workers is the best way to know the fact.[12]

Quality observations require continual learning to develop an appropriate depth of understanding.

Stabilize Processes: *Stability in processes is the bedrock foundation of any improvement system. It creates consistency and repeatability and is the basis for problem identification. Nearly all continuous improvement principles rely on stability because it is the precursor to achieving flow. Many of the rationalizations for waste are based on the instability of processes, as if they are beyond our control. Instead, organizations should apply the basic tools available to reduce or eliminate instability and thus create processes that help to identify and eliminate waste.*[13]

The work of stabilizing processes is to limit the fluctuations within the flow of value. Stabilization requires participation of all levels of an organization. As with each of the supporting concepts, we are again reminded of the interdependence. To reduce fluctuation at the team member level, we expect behaviors that exhibit exact adherence to work standards. Since

---

[11] *Merriam Webster's Collegiate Dictionary*, 11th ed., s.v. "observe."

[12] Ritsuo Shingo, "Go & Observe," MP4 video, 2:54, from an interview by the Shingo Institute, April 2016.

[13] Shingo Institute. *The Shingo Model*, 23.

standards are a component of process that managers owe to team members, we expect behaviors that provide stable operating direction. For example, the operating direction would not vary from quantity to quality based on the delivery cycle or time of month/quarter. At the executive level, we should see consistent behavior in applying all of the Shingo principles.

THE CONSEQUENCES OF NOT FOCUSING ON PROCESS

The following experience demonstrates the potential impact to an organization that is *not* focusing on process.

Some time ago, we were asked to assist an organization that was having trouble completing deliveries on time. The executives believed that the problem existed with the team member who was responsible for scheduling shop floor work. Part of the scheduler's job was to convert the results of the upstream "all electronic" process into hard copy "shop paper." Executives were considering terminating the scheduler's employment.

During a visit to a workplace, we observed that the shop floor appeared to be properly loaded since all machines were running and things seemed to be on schedule on the floor. Suggesting the need for a value stream perspective, we began to lay out the production material and information flow. We discovered that the largest stack of information, both electronic and paper, was in the scheduler's office. But instead of focusing on the scheduler, as the executives had been doing, we focused instead on process by applying scientific inquiry. We asked why there was such a large stack of information in the area. The scheduler told us, "There are so many reasons I can't really list them!"

We determined that the big stack of documents was really multiple smaller stacks, so we asked the scheduler to sort the documents by reason the jobs could not be released to the shop floor. The resulting stacks were sorted into varying categories such as, "engineering has not finished the design," "the customer has not approved the design," "the material is ordered but has not been received yet," and several others. Upon seeing this distribution, the executives asked how an unfinished design (and other reasons) could be on the scheduler's wait-list.

Although the executives did not yet understand, defects were being passed along in the process until they reached the only person in the information flow who could *not* pass them on: the shop floor scheduler. Once we moved upstream and got the executives to focus on process, we discovered that team members upstream were responding to a management edict that was being closely monitored: No one should have a project in their electronic possession (In Box) for more than two days. An attempt to control/optimize/ speed up each work activity had actually created a terrible process.

It is clear from this example how easy it is to blame people, and it's not an uncommon response. In many cases, a bad process can actually appear as if people are creating the problem. Superficial observation of the scheduler's area identified the source of the delayed production, which resulted in placing blame on the scheduler—to the point that he almost lost his job.

In addition, management had been focusing on optimizing operational steps (Shingo's *operations*) versus the flow of value (Shingo's *process*). They reasoned that if team members were encouraged to keep things moving through their stations, production would be completed on schedule. In actuality, this "encouragement" made it acceptable to pass along incomplete (or defective) work. We discovered several violations of principles and supporting concepts in this experience, not the least of which was the shop floor scheduler making things less simple and visual by the way the information inventory was being stored. An interesting value stream revelation to customers, by the way, was that the customer drawing approval process was contributing to late deliveries.

BEHAVIORAL BENCHMARKS

The following behavioral benchmarks—understanding, designing, and attributing—provide a deeper understanding of the principle Focus on Process.

Understand: *We go to where the work happens to develop a thorough understanding of the process.*

We go and observe the process to identify the normal and abnormal conditions that exist. Our observations should concern the flow of value.

Keeping Shingo's words on process in mind—the flow and transformation of material in time and space—will help us to develop the proper perspective.

In order to focus on the flow and transformation of material, we must observe the process. This requires actually laying eyes on what is going on.

If we do not put ourselves in the workplace, we can only guess what is happening in the process. When we are in the workplace, we must keep our perspective on product, service quality, and process quality. Many organizations limit their perspective to results. They rarely ask about process quality or whether or not the organization has provided support for the people who are *in* the process to succeed.

Consistent with Shingo's view, organizations must understand both the process and the operations components.

What does *understand* look like? Ideally, executives and managers will be in the workplace observing the processes simply for the sole purpose of learning. They will not necessarily be providing countermeasures to issues; they will be inquiring about what they see. They might be seeking root cause for problems with knowledge of point of creation versus point of discovery. Without this view, it is impossible to know where to go to take action. Limiting action to the point of discovery can never produce root cause countermeasure implementations. Further, executives may observe behaviors to determine whether they are contributing to ideal results. Managers may observe in the context of the systems and processes provided and may converse in the context of the behaviors the processes drive. Team members will ideally be observed considering beyond their individual work area, particularly considering process steps downstream of themselves, because therein lies the customers of their work. The overall context will be identification of point of creation of issues before pursuing root cause and implementation of permanent countermeasures.

To truly be able to see the process, executives and managers must adopt several of the continuous improvement supporting concepts. The first is to keep things simple and visual. The more complicated the condition of the place where work is done, the more difficult it will be to see reality. Second, they should focus on the entire value stream, which keeps them from missing the impact of what they are observing on the flow of value to the customer. The last two factors, stabilizing processes and creating standard and/or standardized work, are fundamental and they must be actualized before there will be true ability to see and understand the process.

Design: *We design processes to minimize waste.*
Implicit in this behavior is an understanding and utilization of value stream perspective to create end-to-end value streams. This might mean looking at design processes as well as production, quality, service, and delivery processes. Again, Shingo defines process elements as processing, inspection, transportation, and delay. All of these elements apply not only

to production processes but also to development and support processes throughout an organization. The design behavioral benchmark suggests that organizations will consciously eliminate non-value-added activities associated with some of the elements.

Questions to consider that are associated with achieving the design behavior might include:

1. Are drawings clear and presented in such a fashion that minimal effort is required to understand design intent?
2. Is work standardized and presented in such a manner that anyone can comprehend the process with minimal effort?
3. Is work laid out so that everything is in the appropriate location for the team member?
4. Is work laid out to provide appropriate space between process steps?

What else does the design benchmark look like? Behaviors exhibited could include value stream cross functional teams working on design projects. A design effort would ask the purpose of the process, what outcomes would be considered excellent, and what ideal behavior would look like.

Direct observation of organizations over time has taught us that most processes are developed to avoid something, often something that has happened in the past. It isn't often that discussions are about the desired behavior; more often, discussions are about the undesired behavior. If an organization is exhibiting the desired behavior, executives and managers would naturally say, "If this process is working as designed, we will see it when we go to observe the workplace." Design requires executives and managers to start with a broad perspective. Observable behaviors would be discussion of not only the desired outcome as it relates to customer expectation, but also what team members need to be doing on an ongoing basis to sustain excellent customer outcomes.

Reader Challenge

Considering the process you've already begun to review, study and determine the degree to which you can answer affirmatively to the aforementioned questions.

Attribution: *We first look to the process when solving a problem instead of blaming people.*

Exhibiting this behavior in an organization means that seeking root cause is implicit in every investigation. This implies a systematic problem-solving methodology that eliminates people as the cause until all other contributors have been investigated. Another supporting concept, rely on data and facts, is evident in attribution. Exhibiting attribution would create a deep-thinking organization where countermeasures are determined through rigorous analysis of all possible causes as well as the interaction between them. In order to effectively demonstrate this behavior, the organization will utilize information from the supporting concepts.

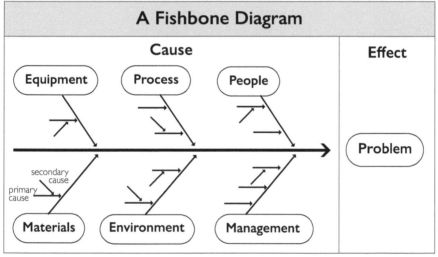

A fishbone diagram is an example of a structured problem-solving method that focuses on relating causes to effect.

What does *attribution* look like? Executives and managers should ask, "What set of value stream conditions created the behavior pattern we are observing?" Managers would not be looking only at operations. There will be evidence that a systematic problem-solving process is being utilized and that executives and managers are teaching others to utilize systematic problem solving.

In addition to a fishbone diagram, other tools that might be used in focusing on the process may include PDCA or some variant, a 5 Whys exercise, Pareto charts, etc.

Reader Challenge

Review each of the Continuous Improvement principles and supporting concepts as discussed in Chapter 2. Then review the previous anecdote, The Consequences of Not Focusing on Process. Identify how the organization realized or violated the Shingo principles and supporting concepts. How does your own organization compare?

6

Assure Quality at the Source

There's no way you're going to reduce defects without processing methods that prevent defects from occurring in the first place.[1]

—**Shigeo Shingo**

BUSINESS CASE FOR ASSURE QUALITY AT THE SOURCE

Assure Quality at the Source is defined by the Shingo Institute. *Perfect quality can only be achieved when every element of work is done right the first time. If a defect occurs in a product or service, it must be detected and corrected at the point and time of its creation.[2]*

FUNDAMENTAL TRUTH

Quality is optimized through ownership and connectedness.

When we put the business case and the fundamental truth together, we may claim, "Because quality is optimized through ownership and connectedness, therefore we must assure quality at the source."

Optimal, according to Shingo, is zero defects. That is the ideal outcome of assuring quality at the source.

---

[1] Shigeo Shingo, *Zero Quality Control: Source Inspection and the Poka-Yoke System*, trans. Andrew P. Dillon (Cambridge, MA: Productivity, 1986), 36.
[2] Shingo Institute, *The Shingo Model*, 23.

DOI: 10.4324/9781003216698-7

THE SUPPORTING CONCEPTS

Two key supporting concepts, Standard Work and No Defect Passed Forward, provide a deeper perspective on the mindset and behaviors needed to assure quality at the source.

Standard Work: *While stability is a necessary precondition for creating flow and improvement, creating standard work builds control into the process itself. Standard work is the supporting principle behind maintaining improvement rather than springing back to preceding practices and results. Standard work also eliminates the need to control operations through cost standards, production targets, or other traditional supervisory methods. When standard work is in place, the work itself serves as the management control mechanism. Supervisors are freer to work on other tasks when there is no need to monitor and control the work process.*[3]

Taiichi Ohno said, "Without standards, there can be no improvement." In fact, where there is no standard, there is also no means to objectively confirm quality. Following is a personal story that illustrates this point.

I was holding a poka-yoke workshop at an aircraft plant and I asked the plant manager if there was an actual mistake we could use for practice. He confirmed that a particular part had not been produced to specification and he supposed that the defect was caused by human error. When we visited the factory to observe, however, the worker was ready for us. He held up two prints, one describing the fixture set-up and the second describing the part. He then asked, "Which of these do you want me to follow, because it's impossible to follow both?"

"Why didn't you tell us about this?" the manager asked.

"I did," the worker replied.

The first lesson from this story is that without a clear standard we will not be able to assure quality at the source. A second lesson is that when defects are assumed to be purely the outcome of human error, we are guilty of ignoring two other key Continuous Improvement principles, Focus on Process and Embrace Scientific Thinking.

No Defect Passed Forward: *This concept is essential for operational excellence from many different points of view. From an executive's perspective, it requires great courage to stop the process long enough to understand the root*

---

[3] Ibid., 24.

cause and take countermeasures that prevent the defect from reoccurring. For the executive, this often means trading any short-term loss for substantial long-term gain. From a manager's perspective, systems must be in place to ensure that any result that varies from the standard, even slightly, creates an expectation of and support for immediate action. This is often called "swarming." From a team member's point of view, "no defect passed forward" requires a mindset of ownership and accountability. If standards are clearly defined, every person should know what good looks like. Executives and managers should role model and then create the conditions for team members to develop the mindset of personal integrity, meaning that no one would ever knowingly forward the outcome of their value contribution to someone else if it contained the slightest variation from the standard. This supporting concept feeds the mindset and tools of continuous improvement and creates the conditions for seeking perfection. It is possible to achieve perfection in the application of this concept.[4]

A personal experience with the concept No Defect Passed Forward offered me a deeper appreciation of its importance:

> About eight years into my organization's Lean journey, my factory foreman, Harvey, approached me one morning with the following proposal: "I'd like your permission to reject outright, any materials at today's Material Review Board that are not absolutely to spec," he said.

With some trepidation, I agreed. Later that day, I was besieged with complaints from other managers. "Has Harvey lost his mind?" they asked. "He rejected every part in MRB this morning!"

When I personally investigated, I was surprised that Harvey had rejected many parts that we would have previously accepted and then sorted or reworked. But I was shocked later when we brought these problems to the attention of our suppliers. "We hadn't heard about this problem," they said.

The consequence to my organization for not understanding the importance of stopping the line for problems was no informative inspection and continuous rework of incoming material.

―――――――――

BEHAVIORAL BENCHMARKS

The following behavioral benchmarks—mistake-proofing, ownership, and connectedness—provide a deeper understanding of this principle.

―――――――――

[4] Ibid., 25.

Mistake-Proofing: *Our processes are designed to prevent, reveal, and immediately resolve every problem.*

People hate making mistakes. We tend to blame ourselves or others for a moment of forgetting or losing track (otherwise known as being human) that could have been avoided by a system designed to prevent the mistake in the first place. Anyone who has driven away from an ATM and left their bank card behind has experienced a process designed to create mistakes. A simple re-ordering of the process that requires the card to be removed *before* the cash is dispensed is an example of a poka-yoke[5] device. Nobody drives away without the cash! Team members love mistake-proof devices and are often the best designers of them. A special creativity and inventiveness comes from 1) an intimate knowledge of the mistake by the persons who made it, and 2) the intrinsic motivation to prevent the mistake from recurring. Mistakes may occur infrequently, but workers worry about those possibilities with every cycle of their work.

In my experience, mistake-proofing is one of the most engaging Lean problem-solving tools. An experience from my consulting provides a good example of mistake-proofing:

Ruthie, an assembler at a control manufacturer, explained to me with frustration that she occasionally forgot to adjust the water pressure during an unattended calibration step for the product.

"When I forget," she related, "I come back an hour later and the product hasn't calibrated."

A reminder sign is not the same as applying poka-yoke thinking.

As we stood by the water valve, I noticed a particularly offensive sign that read: "Don't forget to turn the water off, stupid!"

---

[5] Poka-yoke is a method for trapping human mistakes before they become defects.

I asked Ruthie, "Who would ever put up a sign like that?" She replied, "*I did. I put it up for myself. And I still occasionally forget!*"

We selected her problem as a project for her assembly team. In this case, everyone on her team knew about the problem. In fact, each of them had occasionally forgotten to turn the water off, too. After some brainstorming, one team member noted that he had a quick-disconnect on his garden hose in his yard that served to automatically turn his water on and off. This type of device ultimately solved the problem. No sign was needed. They no longer needed to worry about whether or not the water was turned on. The operation was "poka-yoked" to assure quality at the source.

Ownership: *We design our work so that, at each point in the process, people are able to objectively confirm the quality of their work.*

While this behavioral indicator may seem obvious, considerable historical bias and division of labor have resulted in the responsibility for quality being moved externally to the source. In traditional organizations, ownership is removed from team members. The rationale is that persons cannot objectively check the quality of their own work. Also, while not explicitly stated, this division of labor assumes a Theory X[6] individual working on the frontline—someone who is careless and less intelligent. The inspection function may occur downstream at the next process, or the inspection may be accomplished offline by a separate quality inspection function. In either case, organizations will experience numerous negative consequences:

1. Production will be delayed awaiting the external inspection or, worse, it may continue during inspection, potentially creating more defects.
2. Feedback regarding defects is delayed, which obscures the conditions that caused the defect. Shingo advocated that all inspection should inform the process, providing immediate feedback of a defect to prevent recurrence.
3. Team members are implicitly asked to *do*, but not *check*.
4. Team members often lose even the capability to confirm quality.
5. Loss of control over the quality of their work demoralizes team members and is disrespectful to them.

---

[6] Douglas McGregor's X and Y theories, which were developed during the 1950s and 1960s, are based on motivation in the workplace. Theory X represents a typical employee who is averse to work and Theory Y represents those who are motivated and happy to have an employment opportunity.

PDCA Wheel: Plan, Do, Check, Adjust

In his book, *Zero Quality Control*, Shingo railed on the removal of quality responsibility, from production to the quality department, calling it a "lofty mountain of science inaccessible to ordinary people."[7] While he agreed that people could not be infallibly objective in inspecting their own work, Shingo argued for the development of objective inspection methods and devices that could be deployed at the point of need. In other words, the *check* in PDCA should remain with the *do*. Informative inspection would remove subjectivity from the inspection process and would also provide immediate feedback to the source. The approach gave workers a means to confirm the quality of their processes and to retake ownership of quality; the closer the detection of the defect to the cause, the better it informed the process. Connectivity and ownership were restored, which assured quality at the source. In the case of defects arising from human error, a careful study of the causal relationship between mistake and defect will enable the design of an objective poka-yoke device that could prevent the error from creating a defect. Or, at the least, it could provide immediate feedback after the defect was created and prevent the defect from being passed forward.

The second Insight of the *Shingo Model* declares: *Purpose and Systems Drive Behavior.*[8]

In this case, assuring quality at the source drives both ownership and connectedness, while traditional remote inspection drives lack of attention to quality and detachment from the next process as customer. Perhaps the most unfortunate consequence of *not* assuring quality at the source is that, even if the ownership is removed from team members, they are typically the first to be blamed for defects. The behavior we might expect, as a consequence for being blamed for defects, is *hiding* defects. On the other hand, informative inspection that is managed at the source encourages ownership and connectedness, and as a bonus, respects team members who are the ones most likely to discover the defects. This kind of ownership and

---

[7] Shingo, *Zero Quality Control*, xi.
[8] Shingo Institute, *The Shingo Model,* 10.

control treats defects as gold nuggets to be surfaced by those who are closest to the problem.

It is important to understand the difference between *defects* and *mistakes*. While some defects are created by human error, there are many other potential causes for defects: damage in transport or storage, machine or fixture calibration, incorrect information, improper training, workmanship, or simply lack of a clear standard, to name a few. We see an unfortunate tendency to conflate defects with mistakes, which is disrespectful to team members and may obscure the actual cause of the defect.

As a Shingo Prize examiner, I have visited more than one site where the term *poka-yoke* has been used to describe any objective device that prevents a defect from being passed down to the next process. Most often these objective inspection devices are trapping variation in the process. They are *not* a human error. The consequences of this misunderstanding are that all defects inadvertently imply human error, and they mask the true source of the problem. Such devices are only sorting out defects, not identifying the source of their creation.

Connectedness: *We ensure that people are able to see how they directly impact the lives of others.*

Functional organization and specialization have created many opportunities for miscues and dropped hand-offs. As organizations grow in size, they also naturally grow apart, blurring internal supplier-consumer relationships. As a result, team members at any place in the organization may lose an understanding of the value—to both internal and external customers. In the case of quality, informative inspection is degraded because the identification of defects is so far removed in space and time from the source that the trail has gone cold. On the human side of the equation, interpersonal relationships between internal supplier and consumer can become adversarial.

The following experience in a former job underscores the importance of connectedness. After my company was awarded a Shingo Prize, we became a popular tour site in our region. Students, professors, and businesspeople walked the floor with us hoping to take a few good ideas back home. One team member jokingly referred to it as "Tours R Us." Shortly after we opened our doors for public tours, I was approached by a team member from the welding department.

"You know," Mike said, "the tours are a good idea, but why don't we have some internal tours? I've worked here for more than ten years and I don't know where or how the parts I weld are used."

This seemed like a good idea to me, but I didn't realize its brilliance until our internal consumers and suppliers began meeting face to face. We didn't appreciate the negative consequences of our own lack of connectedness until we saw firsthand its benefits. These benefits ultimately extended to every department as internal tours created broader understanding of each department's impact on the whole.

Shingo railed on the removal of quality responsibility, from production to the quality department, calling it a "lofty mountain of science inaccessible to ordinary people."

A particularly inspiring example comes from Luis C., a mechanical assembler who grew frustrated hearing from a downstream department that his work was defective. The parts he provided in several varieties could easily be assembled backwards, and they occasionally were. Armed with an understanding of mistake-proofing and a clear connection to his customer, Luis took ownership of a defect and devised an ingenious poka-yoke device employing Shingo's methods. This reduced the incidence of defects to zero. As a bonus, Luis mistake-proofed the changeover process for the parts as well and employed quick changeover methods to reduce set-up to "single touch." This enabled the parts to be produced one by one (down from lots of 1,000), as needed.

Ultimately, the fixture and operation were moved to the downstream department, which completed the connection and eliminated the need to schedule and build the part before it was needed. The ideal condition for the principle Improve Flow & Pull had also been achieved for this part, which demonstrates the interconnection between the Shingo Guiding Principles.

One remarkable footnote to this example is that, because the defect identified by Luis C. on the factory floor occurred infrequently, it had previously sailed beneath the radar of the formal AQL system. One-by-one part flow absolutely also requires zero defects and one-by-one problem solving.

Reader Challenges

1. Take a Quality-at-the-Source walk. A posted warning sign such as "Be careful . . ." or "Don't forget . . ." is a red flag for mistakes. These signs are opportunities to design a mistake-proof process. Be sure to engage with the people who work in the department to understand the reason for the sign.
2. Visit for an hour or so with an internal customer of your department. What material or information does your department provide to them? What defects can you find? Identify the root cause and fix, and then confirm it with your internal customer.

KEY SYSTEMS TO ASSURE QUALITY AT THE SOURCE

Most continuous improvement systems have the principle Assure Quality at the Source ingrained in their practice. One of the six rules of kanban, for example, is that "defects are not passed to the next downstream process."[9] However, the following improvement systems are more explicitly related to quality at the source:

1. **Andon:** A visual or audible signal that indicates changing conditions within the factory such as set-up, delays, stock-outs, need for assistance, deviation from schedule, or need for overtime. Such devices are invaluable to bring resources to the point of need when problems occur. When resources swarm to solve a problem signaled by the Andon, it is positive behavior. When the Andon is ignored, this is negative behavior that signals the defects are not important. A secondary indication of bad behavior is an operation where no Andons are alarming. This indicates that management is indifferent to problems.
2. **Stop the Line:** Anyone in the flow of production is authorized to stop the line if a problem occurs in order to assure that the problem

---

[9] Taiichi Ohno, *Toyota Production System: Beyond Large-Scale Production,* trans. Andrew P. Dillon (New York, NY: Productivity, 1988), 30, 41.

is solved at the point where it was created. Traditional organizations have a strong tradition of never stopping to fix. An assembler at an automotive plant related the following to me many years ago: "Our motto is 'Quality Is Job One,' but if I stop the line, they'll fire me." Stopping the line first requires that the team member actually has the capability to confirm the quality of the operation and, second, that they feel empowered and obligated to authorize the stop.

3. **Visual Control:** The practice of visual control systems (VCS) creates a self-managing process where operational status is clear at a glance, often from a great distance. Standardized visual devices enable team members to work without hesitation or delay and to communicate needs quickly. It also helps people identify abnormalities quickly.

4. **Poka-Yoke:** Also known as mistake-proofing, poka-yoke is a method for trapping human mistakes before they become defects. It is a key method in support of Quality at the Source. Shingo's book *Zero Quality Control: Source Inspection and the Poka-Yoke System* is an excellent resource for learning more about it.

5. **Jidoka:** One of the two pillars of the TPS, the jidoka method involves several steps in this Toyota-style automation: the machine detects defects, the machine automatically shuts down the operation when a defect is detected, the machine automatically notifies the operator of the problem. The operator then fixes the problem and re-starts the machine. This is also referred to as *autonomation*.

SUMMARY

Assuring quality at the source requires the combined efforts of executives, managers, and team members to create an environment where objective inspection provides immediate feedback to the process, and where defects are seen as gold nuggets: opportunities for continuous improvement and a professional challenge for all employees—team members, managers, and executives.

7

Improve Flow & Pull

Production plant work and making sushi can be viewed in exactly the same way. Ideally, we want a production system that can maintain producer profits in the face of demands both from customers who want immediate delivery and from those who want to buy only one item.[1]

—Shigeo Shingo

BUSINESS CASE FOR IMPROVE FLOW & PULL

Value for customers is maximized when it is created in response to real demand and at a continuous and uninterrupted flow. Although one-piece flow is the ideal, demand is often distorted between and within organizations. Waste is anything that disrupts the continuous flow of value.[2]

FUNDAMENTAL TRUTH

Eliminating obstacles maximizes value creation.

THE SUPPORTING CONCEPTS

Two key supporting concepts, Identify and Eliminate Waste and Focus on Value Stream, provide a deeper perspective on the mindset and behaviors needed to Improve Flow & Pull.

---

[1] Shingo, *Non-Stock Production*, 41.
[2] Shingo Institute, *The Shingo Model*, 23.

DOI: 10.4324/9781003216698-8

Identify and Eliminate Waste: *Identification and elimination of waste is a practical concept for making processes flow, thus it becomes a primary focus of continuous improvement. Waste elimination is a powerful supporting principle because it is easily understood by everyone associated with a value stream, compared to the complex concepts and computations often associated with cost per unit, cost variances, statistical variability, and other complex metrics. Focusing on the elimination of waste will consistently drive appropriate behavior, while the wrong focus can frequently become a barrier to improvement.*[3]

Waste is anything that slows or interrupts the continuous flow of value to customers. In the end, identifying and eliminating waste is a concept that effectively engages the entire organization in the continuous improvement effort.

Focus on Value Stream: *Improve Flow & Pull combined with Focus on Process necessitates defining value streams and focusing on them. A value stream is the collection of all the necessary steps required to deliver value to the customer. Defining what customers value is an essential step in focusing on the value stream. Clearly understanding the entire value stream, however, is the only way for an organization to improve the value delivered and/ or to improve the process by which it is delivered.*[4]

BEHAVIORAL BENCHMARKS

1. **Uninterrupted:** We design our work toward the continuous creation of value.
2. **Demand:** We produce in response to actual customer demand.
3. **Elimination:** We systematically look for ways to identify and remove waste from our processes.

What is the underlying rationale for the principle Improve Flow & Pull? In my many years as a Lean practitioner, trainer, and coach, I've tried to offer a simple, logical explanation of what continuous improvement is all about. That is, why are we doing this? Figure 7.1 can help us visualize and remember the essence of continuous improvement.

---

[3] Ibid., 25.
[4] Ibid., 24.

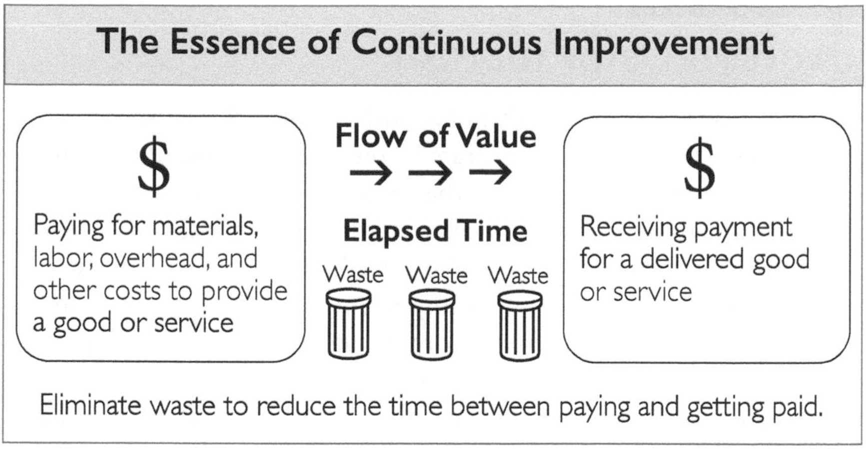

The Essence of Continuous Improvement

Flow of Value
→ → →

$

Paying for materials, labor, overhead, and other costs to provide a good or service

Elapsed Time

Waste   Waste   Waste

$

Receiving payment for a delivered good or service

Eliminate waste to reduce the time between paying and getting paid.

FIGURE 7.1
The essence of continuous improvement is reducing the time between paying and getting paid by eliminating waste.

IMPROVE FLOW

In other words, improvement is focused on creating ideal "flow," allowing the desired product or service to flow rapidly through compact work streams that are designed to operate at the exact rate customer demand warrants. Ideal flow also implies that the product or service is provided without any non-value-added activity included in the processes, *and* without relying on inventory as a buffer to support fulfilling customer demand. This thinking captures the sharp focus needed when pursuing flow improvement.

To elaborate, flow refers to the movement of the product or service through the value stream. Flow is measured by the total elapsed time from start to finish (e.g., order to shipment, concept to launch, phone call answer to complete), where total time for a product or service to be completed includes both value-adding work and non-value-added activities. Flow is increased by identifying and removing anything from the stream that does not add value. Examples of waste, as you may recall from earlier chapters, are things that add time but not value. They could include process and lot delays, transportation, overproduction, and inspection, among other things. As we remove waste from the value stream, we compact the value stream and enable better flow.

IMPROVE PULL, STOP THE PUSH

When we couple the concept of *pulling* with the idea of *flow*, we are adding a focus on customer demand rate to our value stream. Pull focuses on synchronizing and connecting the pieces of the value stream based on the customer's desired rate of consumption for a product or service. In other words, the goal is to operate systems and processes at *exactly* the rate the customer desires. It is important to understand the consequences to customers and to your organization if you do not operate at this rate. Operating the value stream at a rate slower than the customer desires risks not satisfying their needs (e.g., late orders or delayed new product introductions, delayed healthcare appointments, long hold times for customers calling a technical support line). But operating the value stream too quickly implies overproduction, which is considered the worst waste of all.

According to Shingo, products and services produced as needed are "authorized" by the customer because production of the product or service is triggered by an actual order, or a customer pull. In this way, only required items are produced, as opposed to producing items to a forecast. Shingo described producing to forecast as "speculative." It is also referred to as *push*, because items are produced by an upstream process and then pushed downstream whether they are needed or not. Push often results in overproduction and/or unevenness of flow, which leads to many other wastes. Conversely, *pull* production ensures that the right product or service is provided at the right time and in the right amount. It also conserves valuable resources, shortens customer lead time, and increases value to the customer.

Figure 7.2 illustrates how Shingo's thinking compares to more traditional approaches to producing a product. The scenario on the left demonstrates a focus on what appears to be good for the producer. But when you examine it closely, it actually represents processes fraught with waste and stagnation. The scenario on the right focuses on the customer and requires developing operational availability, which is process capabilities that allow a producer to be responsive to customer needs. Embracing *Shingo Model* principles helps us understand how to utilize continuous improvement tools that are focused on improving Flow & Pull. They provide the means to shift systems and processes from the left scenario to the right.

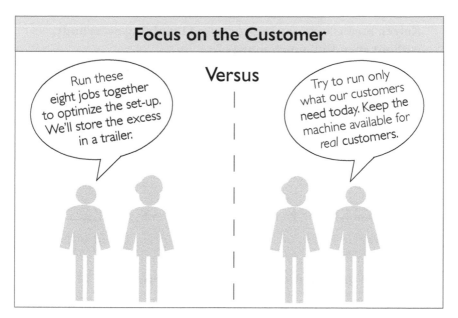

FIGURE 7.2
Develop process capabilities that allow the producer to be responsive to customer needs.

THE GOAL IN IMPROVING FLOW & PULL

Very simply, ideal flow and pull (i.e., one-by-one delivery of product or service without the need for any in-process inventory, delivered at the exact rate the customer desires) is what we aspire to in all our systems and processes. Therefore, pursuing ideal flow and pull means all of our changes and improvements are directed at reducing the wastes that increase the time between paying (for inputs such as material and labor to produce the good or service) and getting paid (delivering the good or service and receiving payment for it). When we focus on this principle, we are challenging ourselves and others daily to identify and then reduce or eliminate all the things that get in the way of flowing value at the customers' desired rate. From product development to material sourcing to marketing to production and delivery of goods and services, the ultimate goal is to reduce or eliminate the impediments to flow in order to deliver value to the customer (both internal and external) when and how they want it.

> **Kaizen means ongoing improvement involving everybody, without spending much money.**[5]
>
> — *Masaaki Imai*

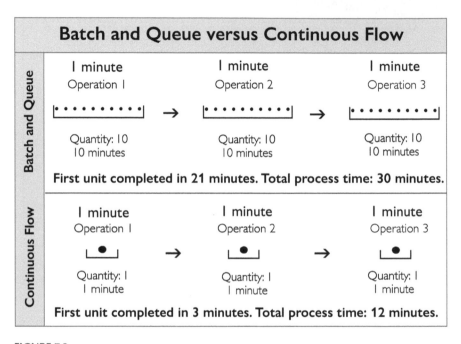

Batch and Queue versus Continuous Flow

Batch and Queue

1 minute	1 minute	1 minute
Operation 1	Operation 2	Operation 3
Quantity: 10	Quantity: 10	Quantity: 10
10 minutes	10 minutes	10 minutes

First unit completed in 21 minutes. Total process time: 30 minutes.

Continuous Flow

1 minute	1 minute	1 minute
Operation 1	Operation 2	Operation 3
Quantity: 1	Quantity: 1	Quantity: 1
1 minute	1 minute	1 minute

First unit completed in 3 minutes. Total process time: 12 minutes.

FIGURE 7.3
An example of batch and queue versus continuous flow.

In traditional organizations, systems often operate in a "batch and queue" manner, pushing work across the value stream. This is the opposite of our desired ideal of systems that flow and pull value. Figure 7.3 helps clarify the implications of a batch-and-queue push system. Let's assume the batch-and-queue push production system requires operational steps 1–3 to build a widget, and each of the three steps takes one minute to transform the item. Currently, those three steps are far apart in the factory. Therefore, widgets being built are moved in batches of ten from one operational step to the next. If we start a new batch of ten in Step 1, when will the first of those ten widgets be completed in Step 3? The answer: 21 minutes

---

[5] Masaaki Imai, *Gemba Kaizen: A Commonsense Approach to a Continuous Improvement Strategy* (New York, NY: McGraw-Hill, 1997), 1.

Step	Operation	Time	Value / Waste
1	Process widget A through step 1.	1 minute	value
2	Wait for nine other widgets to complete step 1. Move all ten to step 2.	9 minutes	waste
3	Process widget A through step 2.	1 minute	value
4	Wait for nine other widgets to complete step 2. Move all ten to step 3.	9 minutes	waste
5	Process widget A through step 3.	1 minute	value
6	Widget A is now complete.		
Total Time: 3 minutes value / 18 minutes waste			

FIGURE 7.4

later, assuming there is no transit time. Why? Figure 7.4 shows the flow for Widget A.

What other potential risks or wastes should we be concerned about in this type of batch-and-queue push system? Here are a few to consider:

1. What if there is transit time added between steps? For example, perhaps a material handler only moves parts every two hours between Steps 1, 2, and 3. *More waiting.*
2. What if a mistake is made at Step 1, and a defect passed but is not caught until Step 3? How many in-process units may be impacted before Step 3 provides feedback on the mistake? *Potential for more defects.*
3. Do we run the risk of damage or loss when we move parts? What happens if we come up short at Step 3? Do we have to set up again and re-make the missing items at the earlier steps once Step 3 reports the shortage? *Potential for more defects.*
4. How much space is needed at each area to store the batch of ten? And what if Step 1 continues to push work to Step 2 even if Step 2 already has a batch of ten to work on? *Waste of storage.*
5. In order to keep track of the batches, perhaps we must transact them on the company computer system when they move from area to area to know where they are. *More non-value-added activity.*

Kaizen is more about using your wisdom than it is about spending money.

— *Shigeo Shingo*

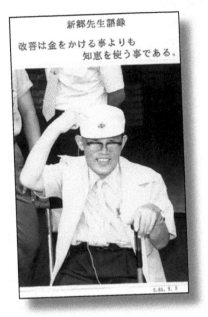

Photo card reproduced and distributed by Shigeo Shingo, dated September 5, 1985. (Original card in possession of the Shingo Institute.)

You should be able to think of even more issues with this type of system.

FLOW PROCESSING AS AN ALTERNATIVE

What would happen if we moved the process steps 1–3 side-by-side and moved the widgets one by one in a continuous flow? And what if one widget is left as beginning inventory at each step at the end of each work shift? Then, when the shift starts, Step 3 finishes its part, then pulls a unit from Step 2, and Step 2 pulls a unit from Step 1? See the new flow in Figure 7.5.

Step	Operation	Time	Value / Waste
1	Complete widget A through step 3.		
	Pull one item from step 2.	1 minute	value
2	Process widget A through step 2.		
	Pull one item from step 1.	1 minute	value
3	Process widget A through step 1.	1 minute	value
Total Process Time: 3 minutes (all value)			
Time to Deliver Widget A to Customer: 1 minute (customer's lead time)			

FIGURE 7.5

And what happens to all the potential risks and wastes we listed earlier when we move to this type of continuous flow using pull? They disappear.

Remember, waste is anything that slows or interrupts the continuous flow of value to customers. Systems pushing work between steps are often discontinuous, poorly synchronized, and not measured against the customer's desired rate of consumption.

DO YOU REALLY UNDERSTAND PULL?

Earlier in this chapter, we discussed flowing and pulling value at the customers' desired rate (takt time). One-piece flow (material, patients, or information, for example) is the *ideal* relative to the flow. Achieving one-piece flow, though not easy and generally attained through a series of kaizen activities over time, requires making changes and applying countermeasures to systems and processes until they can operate in continuous flow. The method organizations use to ensure systems and processes are connected and synchronized at the correct takt time *is* the principle of pull. When pull is in place, no upstream process or operation will provide a good or service until the downstream customer asks for it (or *pulls*). I gave a small example of pull in the second scenario of the widget explanation earlier, but Figure 7.6 is another,

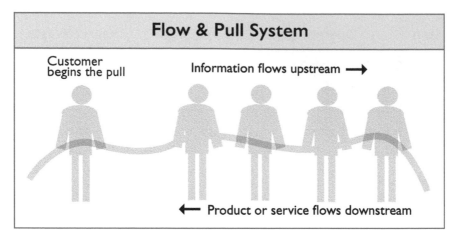

FIGURE 7.6
A flow and pull system causes value to flow downstream by sending information upstream.

more basic, image to confirm your understanding of the difference between push and pull.

Imagine five people standing in a line holding onto a rope. The person at the left end of the rope is the customer who buys a good or service, and the other four people holding the rope represent (from left to right) the parts of an organization that must work together to deliver the good or service the customer on the right desires. Now imagine that whenever the person on the left wants a good or service from the organization, they pull on the rope. When this happens, it authorizes (or sends information to) the person one position upstream of the customer to deliver the good or service. When this person delivers the good or service, they will pull on the rope to signal to the next function upstream that a good or service has been delivered and there is now a need to replenish whatever was used to deliver that good or service. Imagine this pull continuing upstream until the signal gets to the first person in line. Can you see how the rope becomes the unifying link between the customer and the pieces of the system(s) or process(es) that must work together to fulfill the customer's demand?

Connecting and synchronizing the pieces of the value stream that are required to fulfill the demand for a good or service is the basic idea of pull. In the ideal state, the frequency of the pulls on the rope are the same as the customer's desired consumption rate (takt time). Though the concept is easy to understand using the rope imagery, creating a robust pull system in the real world is much more difficult to do. For example, our organizational systems and process steps may be physically separated from one another by buildings, states, or countries. Or, real demand and desired customer rate may be masked by system or process problems, such as large batches of the same type of work being processed together, poor process yield, long changeovers, lack of stable, standardized processes, intermittent delivery patterns, impact of computer system modifiers, and process breakdowns. On top of those issues, our internal policies and structures may be promoting silo behaviors (such as making improvements at a local level that actually decrease overall system efficiency) instead of driving a common focus on customer value delivery. Very often, instead of pulling necessary work through our processes (driving the customer perspective) we are pushing what works for us or what we think the customer wants through our systems and processes. The results, though, are about as predictable as trying to push on a rope.

THE BOTTOM LINE ON IMPROVING FLOW & PULL

The *Shingo Model* sets the expectation that identifying and eliminating waste in pursuit of ideal flow and pull of customer value is a non-negotiable requirement.

Figure 7.7 summarizes key advantages of continuous flow. These are the kinds of outcomes to anticipate when you effectively design your value-creating systems and processes to operate and respond to customer requirements.

Finding and effectively removing waste to enable ideal flow and pull depends on another immensely important resource: people. Respecting individuals is reinforced under the Cultural Enabler dimension of the *Shingo Model*. The key to any continuous improvement effort is effectively engaging the entire organization. The people operating the systems and processes are always in the best position to identify waste and suggest improvements. This implies that leaders be willing to hear team members' concerns and ideas with an open mind and then to work together to make changes for the better. Changes to the systems and processes represent the levers we can move to give team members the ability to drive more value. And through the act of identifying and solving problems, their tacit

Key Advantages of Continuous Flow
✓ Shorter lead times
✓ Less work-in-process inventory
✓ Easier to identify defects and process problems
✓ Keeps team members focused on value-added work
✓ Easier to see where labor is needed
✓ Allows for faster feedback
✓ Improves communication when problems arrive
✓ Reduces many of the seven wastes associated with batch production

FIGURE 7.7
Benefits of continuous flow.

knowledge increases, which further builds the organization's problem-solving and value-delivery capabilities.

As a leader, one of the most important things you can commit to, relative to the principle Improve Flow & Pull, is to embrace your critical role in enabling the removal of waste from your organization's systems. Driving and supporting change in the right direction is the means to ever-increasing customer satisfaction and value delivery, and it is up to you to have enough comprehension of the principle itself as well as Lean system characteristics and tools to understand when and where they can be used to spur improvement. If you don't believe in the principle or understand the role of tools as countermeasures to waste, and you are not clear when it makes sense to use or adjust tools deployed within your business and operational systems, how do you expect your team members to contribute and support this effort? Deploying or adjusting a tool implies change, and team members will need to know why and be comfortable with the rationale for the change in order to embrace it. This is also your job. Barring this, don't be surprised if tools are misapplied, if you fail to achieve what you expect or, worse, the status quo remains, or waste creeps back into the systems and processes.

THE ROLE OF SYSTEMS AND TOOLS

Pursuing ideal flow and pull ultimately means that we are on a never-ending quest to anticipate and provide "JIT means that the right parts needed in assembly reach the assembly line at the time they are needed and only in the quantity needed."[6] Accepting this, it is equally as important to understand the various tools of continuous improvement (such as continuous flow, standardized work, kanban, visual management, set-up reduction, and load leveling) and effectively equip your team members to use them when and where they make sense.

However, executives and managers must beware that tools are simply available as means to counter waste that interrupts the flow of value or gets in the way of pulling work through processes. Remember that the *tools* of continuous improvement are just tools. They are meant to be selected and used when and where they are appropriate to reduce waste and improve the flow & pull of value. Tools are a means to an end, not an end in and of themselves! And like other tools we use in life, the selection of the tool should relate to the problem to be solved. For example, even though a hammer is a common and widely used tool, would you be likely to select it if you had a clogged drain?

Two examples reinforce the difference between focusing on improving flow & pull using tools versus using tools without a proper understanding of their purpose.

A common and powerful countermeasure to establish pull and connect pieces of the value stream that currently cannot be in continuous flow is *kanban*, a signaling method based on pulling work through the value stream by the downstream customer.

During a plant walk-through with a potential client company, I was told they had a kanban system connecting one of the key subassembly areas with final assembly. The intent was to rebuild subassemblies (made by the supplying work center) based on their consumption by the final assembly area (the customer, in this case). I was told the signal to rebuild was the empty bin.

"Okay," I said. "Let's go see."

At the subassembly area I asked the supervisor, "What do you do when you receive an empty kanban bin?"

He answered, "I go onto the computer and check MRP to decide if we should rebuild the box or not."

That told me the supervisor was not following the rules of a true kanban system. It also told me that he did not understand the purpose of the tool that had been deployed in the work area. It is not uncommon to see

---

[6] Taiichi Ohno, *Toyota Production System: Beyond Large-Scale Production* (Portland, OR: Productivity Press, 1988), 4.

this type of scenario with other Lean tools that are misunderstood or have been deployed inappropriately.

The second example comes from the healthcare industry.

A healthcare provider was concerned that one of their clinics did not have enough space in the waiting room. Many patients were standing or queuing up in hallways outside the existing waiting area. The initial response was to request a larger waiting room. Can you see why that traditional countermeasure does not support Lean thinking? Was the problem that there wasn't enough waiting room space or was it really a patient flow problem? After some initial Lean training and involving the clinic staff in identifying value versus waste within their systems and processes, the team experimented with changes to process flow and work area layouts to reduce patient waiting time. In the end, a new, larger waiting area was not needed. Patients were being moved through the process faster and resources were not expended to add more non-value adding (and costly) space.

REAL DEMAND

A key word in the Shingo Institute's definition for Improve Flow & Pull deserves some discussion here. The *Shingo Model* states: *Value for customers is highest when it is created in response to* real [emphasis added] *demand and a continuous and uninterrupted flow. Although one-piece flow is the ideal, often demand is distorted between and within organizations. Waste is anything that disrupts the continuous flow of value.*[7] What does it mean to flow product or service based on *real* demand? And what does it imply about changing to achieve the goal of supplying value based on *true* demand?

Some years ago, I worked for a company that held finished goods in order to meet the desired delivery lead times of its customers. The company also used a computer-based MRP planning system that relied on many planning parameters to determine and suggest when orders to replenish finished goods should be launched. These parameters included things like machine set-up times, sales forecasts, and predetermined lot sizes. Not at all unusual, right? Under this production planning system, we were very

---

[7] Shingo Institute, *The Shingo Model*, 23.

busy making things in the factory and we had plenty of products on the shelf, but we often did not have the *right* items on the shelf to meet customer demand.

How could that be? We were focused on optimizing our production system and believed we were being economical through our planning methods and practices. But, in reality, we were not building to real demand. We were not practicing customer-first thinking, as described in the Shingo Guiding Principle "Create Value for the Customer." Instead, if customers ordered a lot of Product A, we might already be busy for the next few days building Product B while the shelves remained bare of Product A. It was a vicious cycle that was only compounded by things like month-end or quarter-end shipment goals, cost accounting methods that favored large batches due to long changeovers, financial closes that drove period end spikes, and other non-customer-focused behavior.

As we began to understand customer-focused improvement, we set out to improve the flow and flexibility of our production processes by consciously removing waste through the application of tools such as continuous flow cells, standardized work, quick changeover, 5S, and visual systems. We shifted to replenishing finished goods using a visual, rules-based kanban system, rebuilding only what had actually been sold from inventory. We stopped building based on what we thought was going to be sold and what we thought was economical and we made process improvements that allowed us to build to replenish only what had actually shipped. The results spoke for themselves. Customer on-time deliveries were significantly better with less finished inventory on hand. We also had less and less outdated finished goods to deal with and customers were much happier. The team members knew they were building items that were really needed and the old song and dance of second guessing the planners and re-sequencing jobs on the floor came to an end.

As we came to understand that improvement is never ending, and that building to *any level* of finished goods was still not building to true demand, we continued to work with the production and support team members to further improve our systems and processes. It was a quest to become even more flexible to meet true demand. In the end, we made enough improvement to allow the company to eliminate all finished goods. Customer orders for previously stocked items were built and shipped in two days, which was the same promise we had made previously for items shipped from finished goods. (This was back in the day when just getting the paperwork to pull and ship a finished good from the shelf required a two-day promise.)

Reader Challenge

Consider the following questions as they apply to your organization. What changes in thinking are needed in your organization to approach the highest levels of flow and pull? Are you helping team members understand that there is competitive advantage in being able to give customers what they want, when they want it? Do you talk with team members about the wastes associated with inventory (or queued patients or information) stagnating in your processes? How are you helping them to see and remove the system and process issues that get in the way of being able to readily flow work and operate a true pull system?

Talk with team members about flow and pull. What does it mean to them? Do they understand what gets in the way of flow? Can they recognize the difference between push and pull?

THE SUPPORTING CONCEPTS

Focus on the Value Stream: *Improve Flow & Pull combined with Focus on Process necessitates defining value streams and focusing on them. A value stream is the collection of all the necessary steps required to deliver value to the customer. Defining what customers value is an essential step in focusing on the value stream. Clearly understanding the entire value stream, however, is the only way for an organization to improve the value delivered and/ or to improve the process by which it is delivered.*[8]

To better understand and measure the flow of value, I often work with people to develop value stream maps. This important tool, which came to life for many us through the workbook, *Learning to See,*[9] by John Shook and Mike Rother, and the GBMP DVD *Toast: Value Stream Mapping,*[10]

---

[8] Ibid., 24.

[9] Shook and Rother, *Learning to See.*

[10] *Toast: Value Stream Mapping,* directed by Bruce Hamilton (Boston, MA: GBMP, 2009), DVD.

provides a method to capture the current state value stream and envision and lay out plans for its improvement. The current state map provides a high-level image of how well flow is aligned with customer demand (or how distorted the actual flow is from customers' desired flow) and how much of the overall value stream is made of up of things the customer is willing to pay for (which is usually not much in an unimproved value stream).

Interestingly, one of the most basic questions that must be answered in order to build a value stream map is often one of the hardest to get answered in many organizations. What is the rate you have to supply your good or service at in order to satisfy your customers? This is a question about takt time, or the rate at which a customer theoretically consumes a supplied goods or services. For example, how often must we finish seeing a patient so we see all the patients we must we see each day? How often do we need to complete a customer service call so we handle all incoming calls each day? How often do we need to produce a widget so we build enough each shift to satisfy customer demand?

When we think about the ideal flow of value, we always start with the customer in mind. Then the challenge is to develop waste-free systems and processes that are capable of operating at the correct rate (takt time) to satisfy customer demand with the least amount of total inventory across the value stream. In other words, we must keep constant attention on pursuit of the ideal value stream. Yet many people are operating in conditions where they are not clear what the demand is each day for their good or service. Further, they have no clear method to identify whether their systems or processes are currently functioning even close to what was planned, let alone what their customers desire or what is ideal.

Identifying and eliminating waste in pursuit of ideal flow and pull of customer value is a non-negotiable requirement.

For example, when frontline team members are asked, "How do you know if you are ahead or behind?" it is not uncommon to hear them resort to anecdotal evidence. Their replies might be, "the order box is not overflowing," "no one is complaining," "the shelves have plenty of parts on them," or "we don't have too many calls in the queue."

The object is to carry minimal inventories while fulfilling demand for immediate delivery.[11]

—Shigeo Shingo

Executives and managers must acquire important skills like analyzing demand and related patterns, re-calculating takt time, and evaluating and improving value streams. Value stream mapping, when conducted properly, promotes process improvement over operational improvement, and reflects scientific thinking by applying the PDCA method to identifying waste and addressing value stream improvement opportunities. Value stream thinking and documentation methods are essential skills for executives and managers and are critical to understanding what is preventing the organization from achieving ideal flow and pull.

For executives and managers, another key responsibility is to help team members understand what the customer values and then provide team members with the means to measure, understand, and compare their work against the desires of the customer. Tools such as value stream mapping, visual process management, and others are available to help with these efforts. But remember that they need to be understood and correctly deployed to effectively support improvement efforts.

Shingo said, "The most damaging kind of waste is the waste we do not recognize."[12]

He also said,

At Toyota, we have found that there is *always* another way. We look for the waste that people take for granted or don't consider a problem. When we discover a wasteful practice, we don't say, "This can't be helped." Instead, we say, "This doesn't add value, so we'll have to change it—but we'll tolerate it until we can find a way to eliminate the waste completely."[13]

Reader Challenge

Take a gemba walk. Determine whether the processes you observe are ahead or behind. Do team members doing the work know the rate the process needs to operate at (takt) to satisfy customer demand for the day? Do they know how much material and/or information is

---

[11] Ibid., 49.

[12] Shingo, *The Shingo Production Management System*, 35.

[13] Shingo, *A Study of the Toyota Production System*, 80.

supposed to be on hand and why it is there? What do team members do when they are ahead or behind? Is it easy or difficult to see how well the process is doing? What could you do to make the flow rate more apparent and the team members more knowledgeable about the steps to take when the flow rate is too fast or too slow?

Walk your most important value stream starting at the point closest to the customer and then moving upstream through the internal processes. If possible, draw a basic map of the current value stream, like the one shown below, using the appropriate symbology. (The Shingo Award–winning workbook *Learning to See*, by John Shook and Mike Rother, is a great place to learn the symbology and steps to draw a simple value stream map.) Are you able to calculate takt time for your map? Are you able to calculate elapsed lead time and the portion of the lead time that is value added? Where do you see flow disruptions or process stagnation? What are the primary causes for the disruptions?

As you walk your most important value stream, ask the following questions: Are we pulling work through this stream or pushing work through it? Is it all push or all pull or a combination of the two? How are the pieces of the value chain currently connected and synchronized? What happens when problems arise in the value stream? Do we stop and fix, or do team members work around problems?

ENABLING FLOW & PULL THROUGH SYSTEMS AND TOOLS

The *Shingo Model* focuses on developing a culture that reflects, encourages, and supports ideal behavior in order to drive ideal results. The model also purports that *systems* are one of the two key drivers of behavior—either good or bad—within organizations. (The other key driver of behavior is *purpose*.) It follows that to change behavior in the direction of ideal flow and pull organizations will need to adjust their systems. Since executives and managers are typically responsible for the design, management, and control of organizational systems and related policies, these individuals become the owners of the work to evaluate and engage team members to adjust systems under the *Shingo Model*. They must continuously work to align systems with Shingo principles and drive desired behaviors. And since tools are generally embedded within work systems it is imperative to align the rights tools in the right systems in order to drive the right behaviors.

KEY SYSTEMS TO SUPPORT IMPROVING FLOW & PULL

Following are five key systems that closely align with the principle of Improving Flow & Pull.

1. **Improvement Management and Problem-Solving Systems:** Seeing and reducing/eliminating waste are at the heart of improving flow & pull. Improvement management and problem-solving systems must engage all team members in identifying and acting on impediments to the flow and pull of value. In particular, the problem-solving system must provide a structured approach (PDCA) to improvement so that we can hypothesize about and check the results of our improvement experiments. The image in the following related to developing people, which is also found in Chapter 4, Embrace Scientific Thinking, is worth a second look. As executives and managers, we need to put in place systems that help create the need for problem solving, encourage everyone's involvement in the effort, and make a personal commitment to giving team members the time and resources needed to be successful in addressing problems and making improvements. A strong problem-solving system also includes the use of standards to preserve the gains made from successful improvement

Create the Need → Respond to the Need	
Bring Problems to the Surface	**Engage in Problem Solving**
Equipment issues	Solve problems quickly and effectively
Long set-ups	
Rework	Develop team members on the shop floor
New hires	
Unclear instructions	Embrace scientific thinking
Missing parts	Everyone, every minute, every day
Mistakes	Your competitive advantage

FIGURE 7.8
Developing people in support of problem solving.

experiments, such as putting the tool of standardized work in place to ensure repeatable, predictable outcomes every time the work is done. Improvement management systems must set an expectation that the status quo is not acceptable unless we have reached the ideal state of flow and pull. Until then, improvement must be viewed as a normal part of everyone's job, with teams throughout the organization collaborating to improve the work within the systems it occurs in. Executives, in particular, also bear the responsibility of continually raising the bar as gains are made to ensure improvement continues in the direction of ideal.

Standards provide a basis for both maintenance and improvement. Without standards, we have no way of knowing whether we have made improvements or not.[14]

—Masaaki Imai

---

2. **Planning and Scheduling Systems:** Whether your organization sees patients, performs a customer service function, or produces a physical product, your team members' work is subject to the boundaries, methods, tools, and policies provided by your organization's planning and/or scheduling systems. Do these systems currently result in continuous flow and synchronized pull? Do they provide the flexibility to meet changing demand? Do they set an expectation of level-loaded work and operation to takt time? If not, what is being done to adjust these systems to reach these ideals? Are you knowledgeable about and practiced with tools such as cellular layouts, spaghetti diagrams, standardized work forms (such as layouts and playbooks), kanban systems, visual control and management methods, quick changeover, load leveling, and other tools that can assist you in improving flow & pull? Remember, the ideal is producing one by one at takt time to enable your systems to deliver exactly what the customer wants, when they want it, in the quantity they want. Do your current planning and scheduling systems allow you to do this? If not, how will you identify your issues and apply tools to help you move in that direction?

---

[14] Masaaki Imai, *Gemba Kaizen: A Commonsense Approach to a Continuous Improvement Strategy,* 2nd ed. (New York, NY: McGraw-Hill Professional, 2012), 55.

3. **Voice of the Customer:** Value is defined by the customer. Our job is to stay in close touch with customers, both external and internal, to understand what they value. Without a strong system to keep us in close, frequent communication with the customer, how can we be sure that we understand what they value and how well we are meeting their value proposition? How can we evaluate how close we are to achieving the ideal of 100 percent value-added activity at customer takt time? Understanding the voice of the customer is not just a job for executives and managers. Team members who provide the value must also understand customer expectations and accept the challenge of improving their work in pursuit of making value flow faster.

4. **Supplier Management:** In the book *Lean Thinking*, the authors suggest that organizations can only get a quarter to a third of the way on the Lean journey without involving customers and suppliers. Since all value delivery systems rely on inputs such as material or information from some form of supplier, the supplier management system is a critical one to evaluate and improve. Without the right key inputs at the right time, we are sure to suboptimize our value delivery systems.

5. **Visual Management:** The purpose of visual management systems is to make it immediately apparent when something abnormal is happening in a system or process. In the case of flow and pull, visual management systems can be used to signal when flow or pull is not occurring as expected (denoting when something is out of standard). They can also help direct us to the point where the problem started. The goal is to get flow and pull back on track as quickly as possible. Good visual systems support doing this. It may go without saying, but 5S is the natural prerequisite tool for successful implementation of visual management.

The list is not meant to be comprehensive. It just provides a few of the more obvious systems to look at related to flow and pull. There are certainly many other systems and associated tools within your organization that merit close scrutiny as you focus on enterprise-wide flow and pull of value. The *Shingo Model* booklet, downloadable from the Shingo Institute, provides a more comprehensive list of systems for consideration.

SUMMARY

Deep understanding of the principle of Improve Flow & Pull is essential if you plan to achieve the full benefits of the *Shingo Model* within your organization. This principle reflects the basic thinking that we must always be evolving our organizations to effectively and efficiently provide what customers want, when they want it, in the quantity they want. According to Womack and Jones in their book *Lean Thinking*, the goal is "fast-flowing value, defined and then pulled by the customer."[15] Ideally, we would be capable of producing—without generating waste—a good or service in a lead time that is shorter than the customer's consumption interval. When we commit to this thinking, it drives our journey to develop flexible systems and processes that can respond to true demand and react to changes in demand.

Because systems and processes are created and operated by people, they must be engaged in understanding the important whys behind the flow and pull principle, then further engaged, as the experts, in how to adjust the systems and revise the processes and tools to achieve better and better flow and pull. Going to the gemba is the best way to understand the status of flow and pull and to look for the reasons you cannot currently achieve the ideal of waste-free continuous flow. Gemba is also the best place to engage with and check the thinking of those that operate the systems and processes.

As executives and managers, it is our job to regularly see and help others see the current state of flow and pull. We must also create a favorable environment for improving flow & pull through the ideas and creativity of the team members who operate the systems and processes. In the absence of this, organizations risk, as Shingo pointed out, making "improvements aimed at superficial phenomena," while "opportunities for fundamental improvements are continuously overlooked."[16]

Many technical tools are available to assist in the improvement of flow and pull, but it is imperative for executives and managers to understand the purpose of these various tools in order to train and help others to thoughtfully select and apply them as appropriate countermeasures.

---

[15] James P. Womack and Daniel T. Jones, *Lean Thinking: Banish Waste and Create Wealth in Your Corporation*, new ed. (New York, NY: Simon & Schuster, Inc., 2003), 97.
[16] Shingo, *Non-Stock Production*, 23.

However, without appropriate attention to both the social side of improvement (i.e., working with team members in a respectful manner to understand customer value, waste versus work, push versus pull, and the reasons why things are the way they are), your work to implement this critical principle and supporting systems and tools will be difficult and subject to failure. Conversely, people intrinsically want to do valuable work and have amazing ideas on how to remove waste once they can see it. Your job is to objectively see your systems and processes from the point of view of flow and pull of customer value and strive to build a culture that supports and rewards the identification and elimination of anything that gets in the way of ideal flow and pull. Shingo said, "real improvement will not result from superficial imitation of isolated techniques (know-how)" and "new production systems can be undertaken only after a proper understanding of their underlying concepts has been gained."[17]

---

[17] Ibid., 59.

8

Theory in Practice

Following are some practical questions you may ask while evaluating your processes and/or undertaking your continuous improvement activities. We have utilized the supporting concepts associated with the five Continuous Improvement principles in the *Shingo Model* to organize the questions and to give you a deeper sense of what you need to focus on as you study your systems and processes.

When beginning to practice your improvement activities, remember the words of Shingo: "Perceiving means recognizing phenomena by means of our senses. Thinking, on the other hand, is our mental ability to pursue causes and purposes objectively asking *why* about all phenomena."[1]

STABILIZE PROCESSES

The practical implication of stabilizing processes are efforts to eliminate sources of variation. The following questions require a basis for scientific thinking.

Does the machine cycle the same way every time?
Do the parts load the same way every time?
Is the performance board updated at the same time every day?
Is the material being presented to the team member the same way every time?
Are production hours reported at the same time every day?
Are job descriptions formatted the same way every time?

---

[1] Alan Robinson, *Modern Approaches to Manufacturing Improvement: The Shingo System* (Portland, OR: Productivity, 1990), 88. Emphasis added.

DOI: 10.4324/9781003216698-9

Are variations in process captured for analysis?
Are production schedules driven by changeover times?
Are design releases done the same way every time?
Are inventory reconciliations done the same way every time?
Is month-end closing done the same way every time?
Is the production rate repeatable?

STANDARD WORK

Standard work begins with the human interaction component of focusing on process. Providing a set pattern of activity is fundamental to being able to achieve all the other concepts.

Is the description of the work available in the work area?
Do team members refer to the standard work? At what frequency?
Does the standard work include all necessary activities (walking, waiting, etc.)?
Do all support operations have standard work?
Is the standard work verified by observation of the work?
Is the standard followed?
What happens if a deviation from standard is observed?
Are deviations from standard considered to be errors?

GO & OBSERVE

In order to accomplish the Go & Observe concept, team members must have the capability and authority to systematically solve problems. Managers and executives will not be able to observe all occurrences of issues. Only those who do the work can really understand what is going on. Having said that, managers and executives need to be sufficiently present to actually see the process, and the process must be obvious enough for vision to occur very quickly.

Are executives and managers evident in the workspace?
Do executives and managers study what they see or do they act on their perceptions?
Do team members present problems in a consistent fashion?

Do team members present evidence of systematic problem solving (e.g., A3, etc.)?

Is it easy to see the current state of all processes?

Do you see evidence of team members solving problems?

Do Go & Observe activities allow for stopping to study process elements?

When observing Go & Observe activities, do you see a focus on process or operations?

Is the Go & Observe process subject to thinking about improving?

Do we use an observation, idea formulation, judgment, suggestion, implementation improvement model?

Do we look for process improvements before operations improvements?

FOCUS ON VALUE STREAM

We have observed many times that improvement efforts focused on Shingo's operations have resulted in increased disruptions to the flow of value. Quite often, operations are improved in isolation, which results in faster production, fewer quality issues, or other things that are considered good in the context of improvement, but this does nothing to improve the overall delivery of product or service to the customer. Moving product faster through the operation, just to have it sit longer at the next, is not really improvement.

Do you see evidence of value stream maps?

Is the organization aligned with value streams?

Do team members know their upstream suppliers?

Do team members know their downstream customers?

Is it possible for executives and managers to easily see the entire value stream?

Are decisions made only after a value stream discussion?

Is the natural language focused on value streams?

Is storage clearly identified by cause?

KEEP IT SIMPLE & VISUAL

Many organizations have adopted varying criteria that challenge the current condition simply and visually. These criteria include being able to understand what you are looking at from a specified distance in a specified time (say, ten feet in three seconds). Others have adopted time-based determinations of status (one second), problem identification (ten seconds), and countermeasure identification (twenty seconds). Still others utilize 1 (understand purpose of the display), –3 (understand current status), –10 (understand countermeasure that have been utilized). While others suggest that a person should be able to determine who, what, when, where, why, and how within five minutes. The point is to continue challenging the current condition definition of simple and visual to reduce the waste associated with determining what should be managed in the moment.

Is there a standard for how long it should take to determine the current condition of the process?

Can the current condition of the process be determined within the standard?

Can an outsider to the process determine the current condition of the process within the standard?

Have standards for simple and visual been updated?

Is standard work simple and visual?

Can decisions be made based on information presented in gemba?

Is it clear when processing is occurring?

Is it clear when inspection is occurring?

Is it clear when transportation in occurring?

Is it clear when delay is occurring?

IDENTIFY & ELIMINATE WASTE

Best practice begins, in most cases, by asking leaders of the organization why they feel they need to overproduce. Perhaps process and operation instability contributes to this need. Perhaps a subset of instability and product quality creates fear that customer deliveries will be missed. Perhaps the process itself is inherently incapable of consistently producing ideal results. Our observation is that waste reduction is generally focused on operations instead of processes.

Do we emphasize finding waste?

Is inventory considered evil?

Is there extra motion by a team member?

Is defective product being produced?

Are people waiting?

Is product (value) waiting?

How many times is a product picked up and put down before it reaches the customer?

Is it clear when overproduction has occurred?

Do you use value engineering as the first stage in process improvement?

Consider questions you might ask when viewing a process:

Why are the machines arranged the way they are?

Why are batches required?

Why does it take so long to get customer demand to the shop floor?

Are there efforts to eliminate entire wasteful operations via SMED, rearrangement, etc.?

Can you identify internal versus external changeover steps?

When analyzing process, can you identify the set-up operations at every stage in the process?

Can you identify the number of times an object of production is picked up and set down through the process?

Can you identify the amount of after-set-up adjustment?

Can you identify the amount of after-set-up trial runs and adjustment?

Have processes been adjusted as a result of changeover activities?

Is there standard work for changeover operations?

NO DEFECTS PASSED FORWARD

Many organizations utilize Andon systems to implement this concept. Team members can provide visual indicators of problems. Weakness in such systems seem to concentrate in the response given to the signal, and too often the signal is ignored. Leaders must develop a hierarchy of response successively involving levels of management who are able to respond and implement countermeasures. The best systems we've seen have an information distribution system (such as mobile phones or broadcast announcements), which escalates contact based on the time passed since the trouble signal was initiated.

Are there quick feedback loops?
Are poka-yoke efforts evident?
Are poka-yoke efforts clearly defined as either control type or warning type?
Can team members stop the line when defects occur?
Are incomplete documents considered to be defects?
Can team members stop the process when errors occur?
Are errors considered to be different from defects?
Have operations identified input requirements for their operation?
Do operations know output requirements to satisfy the next operation?
Are you replacing judgment inspection with informative inspection?
Do you distinguish between point-of-error creation and point-of-discovery creation?
Do you distinguish between isolated defects and serial defects?
Do you utilize sensory inspection or physical inspection?
Do you use subjective inspection or objective inspection?
Do you utilize process-internal inspections?

INTEGRATE IMPROVEMENT WITH WORK

If integrating improvement is going to be accomplished, a fair measure of the other concepts must be embedded in the flow of value. If it's not embedded, thinking about improvement, for executives, managers, and team members, is an additional burden on top of running the operation and completing daily work requirements.

Is time allowed for improvement every day?
Are there team and individual improvement performance metrics?
Do team members actually utilize the time allowed for improvement?
Are there simple methods to capture improvement ideas?
Do improvement ideas require review?
Is it easy for team members to call out waste?
Is it easy for team members to modify operations to eliminate waste?
Is it easy for team members to modify processes to eliminate waste?

RELY ON FACTS & DATA

The reality of this concept is the measurement, by some mechanism, of results, condition, and behaviors. To fit the definitions, one must be able to say, "I have seen it, the object, and it has occurred this number of times." This can relate to production results, problem definition, or observed behaviors.

Is there statistical process control (SPC)?

Are there consistent efforts to eliminate the need for SPC?

Is there a clear definition of normal?

Do executives and managers go to where events occur to discuss them?

Can executives and managers state that they have seen events occur?

9

Improvement Systems and Behaviors

Management must do Kaizen, too.[1]

—Hajima Oba, General Manager, Toyota Production
System Support Center

Since the early 1980s, many organizations have invested countless hours in training for improvement, selecting from a menu that typically begins with a Lean 101 primer about the seven wastes and then continues through a broad range of technical improvement subsystems, each designed to reduce waste and improve flow. A short list typically includes these tools:

1. **5S:** Workplace organization to create order and stability.
2. **VCS:** Visual control systems to make conditions clear at a glance.
3. **S/W:** Standardized work to systematically document the best current combination of team members, machines, and material for a given level of demand.
4. **P/S:** Problem solving, which has a broad range of subsystems to diagnose and remediate problems, such as 5 Why root-cause analysis, improvement kata, six-step problem solving, DMAIC, 8D, A3, and mistake-proofing.
5. **TPM:** Total productive maintenance of equipment to assure consistent quality and rate.
6. **SMED:** Single-minute exchange of dies (or quick changeover) to make it easier to produce the customer's exact order at the right time.
7. **Continuous Flow:** Physical organization of resources to facilitate completion of all production operations in a single location with minimal waste and delay.
8. **Value Stream Mapping:** Examining and improving the flow of material and information within a defined system.

---

[1] Hajima Oba, conversation with Bruce Hamilton, 1998.

DOI: 10.4324/9781003216698-10

Three Insights of Organizational Excellence
1. Ideal results require ideal behavior.
2. Purpose and systems drive behavior.
3. Principles inform ideal behavior.

FIGURE 9.1
Three Insights of the *Shingo Model*

Over time, the menu of tools has grown much longer, but too often, the results are not compelling. After many years of working with organizations large and small across many different industries, we have concluded that a formulaic technical approach to operational excellence degrades quickly if it is not grounded in foundational principles.

As noted by the Shingo Institute's Three Insights, *ideal results require ideal behavior*,[2] and that behavior is a product of the right systems, grounded in fundamental principles. Quality and productivity tools are necessary for continuous improvement, but they are not sufficient. In fact, when the tools and systems of continuous improvement are simply layered over a traditional, autocratic business culture, the results are frequently disrespectful of people. For example, a 5S project may be executed without input from the very persons who work in the area to be organized. Or, standardized work is approached merely as a time-setting exercise performed by engineers with stopwatches. Improvement systems that are originally intended to support and engage workers instead become disengaging and demoralizing. It is little wonder that these systems do not sustain. Improvement systems that are not informed by principles encourage behaviors that are far from ideal.

But what are those ideal behaviors? We are often asked that question when we present Shingo Institute workshops. In the context of improvement systems, what do ideal behaviors look like? More specifically, because these systems are the primary focus for managers, what do ideal *manager* behaviors look like?

Bruce explains:

> In 1998, my teacher, Hajime Oba, who had been providing assistance to my company for about four years, gave me an assignment to engage in my own

---

[2] Shingo Institute, *The Shingo Model*, 10.

Kaizen. "Management must do Kaizen too," he said. He didn't explain why, and I didn't ask, but I assumed it was to get me closer to the floor. With the help of some folks from production, I created a list of improvements to work on and posted a log of what I had done. Some ideas were better than others and some generated a few laughs from team members. Along the way, I learned that even small changes can be uncomfortable, but also that small changes are small only in cost of resources, not in their effects. I think I had understood this at some level, but direct participation is a good teacher. It helps a manager to respect every individual and to lead with humility. But the most significant learning for me was that my personal engagement with Kaizen was a thousand times more impactful in creating a culture of continuous improvement than any amount of "atta boys" I could have bestowed on others for their improvement efforts. I believe this was the key point the Mr. Oba wanted me to take from the experience: that each of us models our behavior according to those to whom we report. Ideal behavior from the boss begets ideal behaviors in team members.

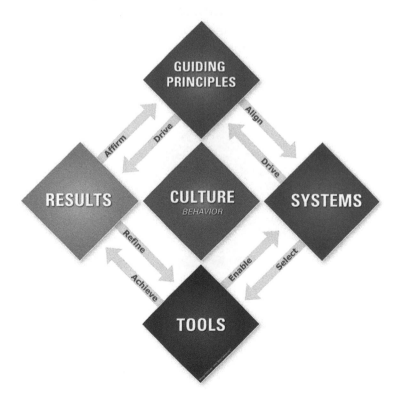

FIGURE 9.2
The Shingo Institute suggests executives, managers, and team members invest 80 percent of time and focus on improvement. Everybody, every day.

GBMP's 2011 Shingo Award–winning DVD, *Moments of Truth*,[3] addresses the critical chain of support from executives to managers and frontline supervisors that is needed to engage *all* team members. As we say at GBMP, "Everybody, every day." And as Shingo examiners, we are keenly observant of the behaviors of managers in relation to the continuous improvement systems they provide to team members. Do manager behaviors create a favorable environment, or do they stifle and extinguish team member engagement?

Ideal manager behaviors rely first on a clear understanding of the concepts behind the improvement tools. Shingo always advocated that we should first know why, then know how.[4] Managers cannot provide passionate support for tools they do not understand. Second, the passionate support must also be guided by the Shingo Guiding Principles. For example, a manager may understand the mechanics of root-cause analysis, but if they are inclined to blame team members for problems, the principle Focus on Process is violated and team members will hide problems. An arsenal of problem-solving tools is useless if managers punish team members who report problems.

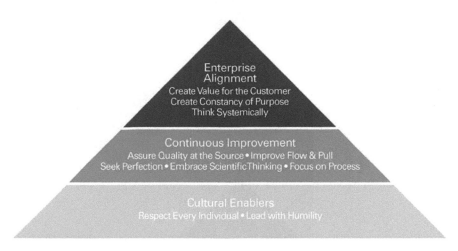

FIGURE 9.3
The Shingo Guiding Principles

---

[3] *Moments of Truth*, directed by Bruce Hamilton (Boston, MA: GBMP, 2009), DVD, www.gbmp-streaming.org/programs/moments-of-truth.

[4] Shingo expressed this concept numerous times in many different ways. For example, see Shingo, *Non-Stock Production*, 59.

WHAT DOES IDEAL LOOK LIKE?

But how do we assess behaviors as they relate to continuous improvement systems if there is no ideal image as a standard? What does *ideal* look like? To provide some guidance in this regard, following is a list of critical subsystems that impact behaviors at all levels of the organization. We have grouped them into five broad categories:

1. **Management:** These are systems that correspond roughly to the Enterprise Alignment principles as they relate to providing high-level guidance for continuous improvement.
2. **Engagement:** People development systems are critical to providing both the information and inspiration to team members. Fostering engagement requires special remedial attention to overcome a traditional culture where most team members are not included in problem solving and improvement. This category aligns with the Cultural Enabler principles.
3–5. **Stability, Standardization, and JIT:** These three categories of improvement systems and subsystems, which are typically referred to as Lean tools, describe the journey that organizations must follow to transform from traditional methods to operational excellence methods. These systems correspond to the five Continuous Improvement principles in the *Shingo Model* that are described in this book.

When assessing applicants for the Shingo Prize, Shingo examiners look deep into organizations to identify the Shingo Guiding Principles in action through the behaviors of the people and observe whether or not systems are effectively driving ideal behaviors.

THE SHINGO BEHAVIORAL ASSESSMENT SCALE

The difference between observed behavior and ideal behavior is the behavioral *gap*. The Shingo Behavioral Assessment Scale, Figure 9.4, provides a gap analysis that can be used to determine where and how you will begin your improvement activities.

BEHAVIORAL ASSESSMENT SCALE

This list of descriptors is the basis for assessing behaviors in an organization. Behaviors that match the descriptors would score at the top of the indicated range.

Lenses	Level 1: 0-20%	Level 2: 21-40%	Level 3: 41-60%	Level 4: 61-80%	Level 5: 81-100%
Role	Executives are mostly focused on fire-fighting and largely absent from improvement efforts.	Executives are aware of others' initiatives to improve but largely uninvolved.	Executives set direction for improvement and support efforts of others.	Executives are involved in improvement efforts and support the alignment of principles of operational excellence with systems.	Executives are focused on ensuring the principles of organizational excellence are driven deeply into the culture and regularly assessed for improvement.
	Managers are oriented toward getting results "at all costs".	Managers mostly look to specialists to create improvement through project orientation.	Managers are involved in developing systems and helping others use tools effectively.	Managers focus on driving behaviors through the design of systems.	Managers are primarily focused on continuously improving systems to drive behavior more closely aligned with principles of organizational excellence.
	Team members focus on doing their jobs and are largely treated like an expense.	Team members are occasionally asked to participate on an improvement team usually led by someone outside their natural work team.	Team members are trained and participate in improvement projects.	Team members are involved every day in using tools to drive continuous improvement in their own areas of responsibility.	Team members understand principles, "the why" behind the tools, and are leaders for improving not only their own work systems but also others within their value stream.
Frequency	Infrequent • Rare	Event-based • Irregular	Frequent • Common	Consistent • Predominant	Constant • Uniform
Duration	Initiated • Undeveloped	Experimental • Formative	Repeatable • Predictable	Established • Stable	Culturally Ingrained • Mature
Intensity	Apathetic • Indifferent	Apparent • Individual Commitment	Moderate • Local Commitment	Persistent • Wide Commitment	Tenacious • Full Commitment
Scope	Isolated • Point Solution	Silos • Internal Value Stream	Predominantly Operations • Functional Value Stream	Multiple Business Processes • Integrated Value Stream	Enterprise-wide • Extended Value Stream

FIGURE 9.4

The Shingo Behavioral Assessment Scale

All levels of an organization—executives, managers, and team members—can evaluate their behavior using this scale. You can also use it to help you determine how deeply embedded a particular behavior is in your organization's culture. Keep in mind that both good and bad behaviors can be embedded. To assess how deeply embedded a behavior is in your organization's culture, ask these five simple questions:

1. How often do we see the behavior? (Frequency)
2. Are we seeing the behavior for the first time or have we seen this behavior for years? (Duration)
3. Is there a sense of passion and importance for the behavior? (Intensity)
4. Do we see the behavior in just a few areas or is it widespread throughout the organization? (Scope)
5. Who exhibits the behavior—executive, manager, or team member— and is it exhibited at the appropriate level? (Role)

We include the Shingo Behavioral Assessment Scale here to show *what good looks like*. Much of the value of the Shingo Institute workshops derives from the self-reflections of workshop participants. The scale provides an additional means for self-assessment that will provide a concrete standard for observation. Remember, benchmark behaviors are not a checklist, nor do they provide a solution to behavioral problems.

To analyze the scores from the Behavioral Assessment Scale, Shingo workshops attendees participate in Go & Observe activities and then write specific observations on post-it notes and place the notes in the appropriate "swim lane" (the vertical lines on the Shingo Behavioral Assessment Scale). Anything that is assessed to be less than Level 5 demonstrates a gap between observed behavior and ideal behavior. When participating in swim lane scoring, participants follow the standard work of examiners when they are evaluating a facility.

While the Shingo Institute utilizes the "swim lane" tool to visually identify gaps in behavior, other visual tools, such as radar charts, can also be used effectively to accomplish the same effect.

As mentioned earlier, these visual assessment tools can be used to identify behavioral gaps not only on an organizational level, but also on a personal level. In other words, you can use these to evaluate yourselves.

Reader Challenge

Use the Shingo Behavioral Assessment Scale to determine how closely your behaviors match the descriptors. Then analyze your score by drawing the basic outline of Figure 9.5 on a sheet of paper (only include the vertical lines). Any behavior that is assessed to be less than Level 5 demonstrates a "gap" between observed behavior and ideal behavior. See Figure 9.5 as an example.

Observed Behavior versus Ideal Behavior					
	Level 1	Level 2	Level 3	Level 4	Level 5
Role					X
Frequency				X	
Duration					X
Intensity					X
Scope			X		

FIGURE 9.5
"Swim lanes" help to easily recognize achievement gaps.

SUMMARY

While the focus on change management is frequently on frontline engagement, the single most significant step toward organizational excellence can be found in engagement of top management, middle managers, and first-line supervision. Ongoing self-reflection and development at these levels will ensure the creation of a continuous improvement culture.

10

Where Do I Go from Here?

We start with this first question: "What is the business problem we must solve?" Then we apply lean tools and principles to create a model cell. Again, the key is tying that work to the business goals of the organization. This model cell experiment—tied to key performance targets—creates a new system with standard work that sustains improvement and aligns everyone on larger organizational goals.[1]

—John Shook

In this chapter, we will share practical knowledge on how to leverage what you have learned from the prior chapters in this book. In other words, this is a short primer on how an organization can begin to implement the principles, systems, and tools encompassed in the *Shingo Model*'s Continuous Improvement dimension.

Our intent is to give you a proven framework, endorsed and practiced by successful Lean organizations such as Toyota, for getting started (or re-started) on your Shingo-based journey. This method respects the needs of your customers, your workforce, and your organization. It also reflects the reality that the CI dimension of the *Shingo Model* relies on concurrent application of the other *two Shingo Model* dimensions (Cultural Enablers and Enterprise Alignment). Most importantly, it offers the means to experiment and learn in a safe, small space as you get started. The seven steps given in the following will help you begin or further your continuous improvement journey and put your knowledge into action.

Before proceeding, it is worth reinforcing once again the fundamental thinking from Shingo about the integrated nature of improvement. While he is most often remembered for introducing technical concepts like quick changeover and mistake-proofing, which are key tools to enable systems

---

[1] John Shook, speech. Lean Transformation Summit, Orlando, FL, March 5, 2014.

DOI: 10.4324/9781003216698-11

to increase the flow of value, we believe Shingo's greatest contribution was providing an integrated image of Lean. This is a system that relies on both technical and social science aspects—tools *and* culture. One cannot exist without the other. Your pilot efforts at applying continuous improvement principles and methods *must* focus on the development of both sciences.

Your learning during this phase will inform your future continuous improvement journey. Furthermore, your thoughtful consideration of how to enable both the social and technical aspects of Lean from the outset is what sets the stage for your organization to achieve the full benefits of the *Shingo Model* over time.

WHAT IS YOUR CURRENT CONDITION?

In Chapter 9, we introduced a method to conduct a system assessment. Whether you use this or some other method, you must have a *true* understanding of your current condition before you begin your improvement journey. Remember that continuous improvement is about making changes in the right direction and, in the case of the *Shingo Model*, according to the ten principles and supporting concepts.

A word to the wise as you go about assessing your current condition and considering what could be: We are all subject to conceptual blind spots when evaluating our own organizations and considering what can be accomplished. Following is a story to help you understand conceptual blind spots.

In 1996, a consultant from a division of Toyota began working with some of us on a changeover reduction project in our plant's machining area. This internal supplier was a key provider of final assembly parts and was a common source of parts shortages negatively impacting final assembly. The consultant was adamant in pressing us to find ways to get to under ten-minute changeovers on the pilot machine and from the outset we had doubts it could be done. However, it quickly became apparent to us through her questioning that our doubts arose from "conceptual blind spots" developed through old set-up practices. Our history, experience, and closeness to the process were preventing us from seeing a multitude of improvement opportunities. With some priming from our consultant, we found waste in areas where Shingo would say "it was thought not to exist."[2] Under the guidance of our TSSC consultant, our well-ingrained

---

[2] See, for example, Robinson, *Modern Approaches to Manufacturing Improvement*, 79.

existing frameworks for analyzing problems were gradually *un-learned* as part of our Lean learning. With blinders removed we were able to get the changeovers down to eight minutes.

How does this relate to your initial system assessment? If you and your organization don't know what ideal looks like, you will not be able to see the reality of the gap between where you are and where you need to head—the ideal. This is why you have to commit to deepening your own learning. As a leader, you must not only see the gaps yourself but also have a responsibility to help others see and understand the gaps. Reading, visiting other companies, attending workshops and conferences, personally participating in kaizen activities, drawing value stream maps, attending and/or hosting Shingo training, and getting a Lean mentor or coach are some ways you and others in your organization can deepen your learning.

So, the first step in getting started with your continuous improvement implementation is to commit to clearly understanding the current state in your organization. A good Lean system assessment tool provides a great way to ensure that your team is looking at the right things—trying to understand where you are and what the pathway to ideal might look like. We encourage you to work as a team to document your current state, regardless the method you use, as you will be surprised how quickly you and others forget where you were, how you assessed your gaps, what key areas you assessed looked like at the outset, etc.

Ideally, you should reach consensus among the leadership team of your organization regarding the results of the system assessment. You may need to do some initial training with the group before completing the assessment. You may also want to consider using an external facilitator to help with the assessment so you have an outside set of eyes to guide you. They would have deeper experience and fewer conceptual blind spots. At the end of your initial assessment, a key goal should be to have greater alignment among your team relative to where you are and the size of the gap between your current state and an ideal Lean implementation.

WHAT IS YOUR BURNING PLATFORM?

With your system assessment complete, the next step is to develop consensus around what the most important improvement focus should be. This might come from a high-level discussion about the most important things your organization needs to work on to become more successful or to sustain your success. For example, when was the last time you had a specific discussion about the value you deliver to your customers, the state of the systems and processes you use to deliver this value, and how you measure the degree of success you are achieving in delivering this value? The intent of the discussion is to uncover the most important area of improvement focus—your burning platform, if you will. What do you need to focus on today to place you, or keep you, on the path to sustaining long-term success of the organization?

Here's an example to help you understand what a burning platform might look like in the real world. Company A, one of many competitors in a segment of the sporting goods arena, determined that customers were interested in having new and different products more frequently. Many customers owned several products *and* were regularly looking to buy the latest items on the market. Yet Company A's product design and release process was long and siloed, and it was costly due to the long development cycles and a plethora of product problems that had to be fixed after launch. Challenged by their leader who had significant experience with continuous improvement in his previous roles, they decided that their burning platform was the new product development system. They agreed to focus on increasing the flow of new, high-quality products to the market. This led them to a project to significantly improve their product to launch value stream. Using Lean principles, systems, and tools, they focused on inhibitors to flow, eliminating waste, and promoting close customer/supplier relationships between everyone who played a part in the design to launch process. These efforts touched many functions and systems and literally eliminated walls between people who worked on aspects of the new product process. Their results have been impressive and the anecdotal evidence of the positive impact on team members involved in the new system is even more exciting. They talk about the many changes that have allowed them to stay focused on creating exciting new products that can be manufactured and delivered to customers from day one! Their time to market for new products has been reduced by more than 50 percent and their measure for revenue from new products has gone up dramatically over the last few years.

WHAT ARE YOUR ACCELERATORS
AND INHIBITORS?

Now that you have a realistic understanding of your current condition and have agreed on your burning platform, the next important step is to identify existing conditions within your organization that either accelerate or inhibit your improvement efforts. The intent here is to leverage the accelerators and manage and/or minimize the inhibitors. It makes sense to engage a group to develop this list of conditions so that you look widely and avoid any potential conceptual blind spots. Following are some items for your consideration as you carry out your accelerator/inhibitor analysis.

Depth and Breadth of Existing CI Efforts

If you are already successfully engaging many people and areas in some form of improvement it will likely be easier to staff and succeed with a deep pilot effort. This doesn't mean you won't need to talk to people about why additional changes are needed and it doesn't mean there won't be obstacles, but it does mean that people may already have a foundation to accept additional change. Conversely, if this is your first run at continuous improvement, you will need to spend more time upfront to promote and explain the efforts, plan for ongoing communications, and gain trust and alignment.

Existing Management System to Support CI

This refers to looking at the CI support methods you already have and comparing them to what you will need. For example, if executives and managers currently don't spend time regularly in the gemba, interacting and checking on their systems and processes, this is an inhibitor. Or, if you currently don't have access to knowledgeable resources to teach and coach in Lean principles, methods, and tools this will make your initial efforts much harder. If your management style has historically been "command and control" as opposed to inclusive and open to all ideas, this is an inhibitor. If time devoted to improvement is seen as wasteful, this is a problem.

Reporting Structure

Another part of reviewing your management systems relates to looking at your existing reporting structure to evaluate how well equipped you will be to provide the training, coaching, and day-to-day support needed for CI to flourish. For example, if we looked at an organizational chart for your organization would the reporting relationships be one to many or something like one to five? In the book *Toyota Culture*, the authors point out that teamwork is a fundamental element and the perfect number of team members on a team is five. They believe that small teams are easier to control and more capable of solving problems and maintaining mutual trust. The authors also point out the following three attributes of the Toyota organizational structure: 1) use of matrix style of management, 2) the distribution of decision-making policy is based on exact standards, and 3) clear identification of the team leader's functions.[3]

Current Policies and Procedures That Support CI

In our experience, everything is subject to review and change over time in an effective Shingo-based improvement journey. However, existing policies and procedures codify the status quo. Therefore, in most organizations leaders and managers—generally the owners and keepers of these tools—must be careful to understand whether the behaviors driven by these policies and procedures will inspire or detract from improvement efforts.

Here is a scenario to illustrate this concept. The procedure to order parts from a supplier required a signed requisition that then had to be signed by the department manager, who then sent it to the commodity manager. The commodity manager then created a purchase order which then had to be approved and signed off by the purchasing manager. This was the policy whether the parts cost $0.00001 or fifty dollars each. During an improvement event to address purchased part lead times a team determined they could save five days on average if this procedure was altered to allow the warehouse staff to signal a kanban release directly with the supplier. The release would go against a blanket purchase order that the purchasing group put in place. However, there needed to be support and willingness to change the long-established purchasing procedures to allow

---

[3] Jeffrey Liker and Michael Hoseus, *Toyota Culture: The Heart and Soul of the Toyota Way* (Columbus: McGraw-Hill Education, 2008), 235.

this important change. Our experience is that these types of policies and procedures are not unusual in many areas in organizations. The average team member has to live by them but needs the help and support of managers and executives to change them.

Measurements That Encourage CI

Most of us are familiar with the phrase "measure what matters." In the *Shingo Model*, this ultimately should drive us to measure what matters to the customer (value) and what matters to team members (not having any obstacles in their efforts to provide value). We see many organizations who have dozens of key metrics in place, but most are lagging metrics and they don't have corresponding leading indicators to drive meaningful, real-time improvement efforts that will ensure the most important customer and team member metrics are achieved. Worse yet, many team members can't relate to the existing measures and/or speak to how they can impact those measures. You might hear statements like, "Those numbers are for managers. They don't really mean anything to me."

In some cases, existing measures may even inspire us to change in the wrong direction or preserve the status quo. Let's consider one example that some of us have experienced. After freeing up capacity from improvement efforts, some sub-contracted work was brought back into the plant. In other words, we had existing space and team members available to do the work so we could gain control of the lead time, quality, and inventory levels associated with these products. However, the existing cost accounting systems and related policies on how cost burdens were applied suggested the margins on these products would now be unacceptable if they were made internally, much to the dismay of the product manager. Fortunately, the CFO could see beyond the numbers in the cost accounting system and explained that it made sense to use the free space, equipment, and people to bring the work inside. It was the costing system that was wrong, not the thinking.

Shigeo Shingo advised that "99 percent of objection is cautionary."[4] In other words, those who appear to vigorously object to Lean are really just asking for more information. In examining your accelerators and detractors, you need to evaluate what behaviors you are currently driving and what messages you are reinforcing about importance of change and improvement in your organization.

---

[4] Shingo, *Non-Stock Production*, 182.

WHAT IS YOUR PROOF-OF-CONCEPT PROJECT?

The phrase "a mile wide and an inch deep" is probably more familiar than "an inch wide and a mile deep" to you. Attributed to Edgar Nye, an American journalist and humorist, the better-known phrase is said to have been coined by Nye when he was describing the Platte River in 1899, a significant tributary to the Mississippi River. The Platte is a muddy, wide, shallow, meandering river and hence it was not deemed navigable. It was considered of little utility or influence at the time. Subsequently, the phrase was adopted to describe people, academia, and other fields or situations where knowledge of things is superficial.

When we apply the "mile wide, inch deep" phrase to a CI implementation, we are referring to cases where some amount of Lean training or practice has been widely broadcast across an organization, but with little real depth of learning relative to principles, systems, and tools. And often, despite some early gains, improvements have not sustained over time and the status quo has crept back in.

For example, perhaps everyone got a two-hour introduction to Lean and the organization has done a bunch of 5S activities all over the site. Maybe initially the staff and visitors noted the place looked neater and there was some excitement that things could change. Yet, a month or two later, things looked exactly the same as before the training and 5S events took place. On top of that, no actual improvements to the flow of value had been accomplished for the time and effort invested, team members are performing work much the same, and worse yet, many have written off Lean as just another "program of the month" that had come and gone with no real benefit to them, the customers, or the organization.

When we flip the thinking to promote continuous improvement based on "an inch wide and a mile deep," we are suggesting that you start by focusing on a narrow portion of your business and do a deep implementation in a single pilot area, system, or process. The idea is to structure the initial improvement project as a controlled experiment that is consciously constructed to demonstrate how the technical and social sciences of Lean work together to elicit real, sustainable improvement. The pilot should focus on systematically improving the flow of value in ways that reinforce Lean principles, includes appropriate management oversight and support systems and policies, applies tools appropriate to the problems that need to be countered, and demonstrates how to show respect for those involved in or impacted by the pilot. The pilot approach also serves in a small way to show the rest of the organization the steps necessary to envision and achieve the full benefits of a principle-based implementation. It provides a structured learning environment for early participants and allows those outside the pilot to observe the experiment from afar. The intent is for the pilot to pave the way for broader and deeper CI efforts after its completion. An early success through a pilot helps deepen the understanding and acceptance for both the technical and social aspects of Lean.

The choice of area, system, or process for your pilot should be well thought out. In fact, you should stack the decks in your favor for an early success. What might this mean? Here are some ideas:

1. **Pick a pilot in areas, systems, or processes** that are overseen by an executive and/or managers who are supportive of and interested in changing the work for the better. Perhaps they feel pain from the targeted systems or processes and have a personal desire to decrease the pain. This means they will be more invested in the pilot effort.

2. **Look for places where there is a need,** from the point of view of the customer, the team members, or the business. For example, you may have a product or service line where demand is growing, and you are having difficulty meeting the demand with existing systems, processes, and people. Or you may have a system that is producing defects that are getting to downstream customers. Improvement should always be based on first asking, "What is the need?"

3. **Pick an area, system, or process** where there is a natural beginning and end. It should have boundaries that are largely within the local span of control (as opposed to one that covers a worldwide process and involves many people or locations who must agree to/sign off on any proposed improvements).

4. **Pick a pilot where you can fairly easily document the current state,** walk the flow, and gather good data to understand current processes, operations, metrics, trends, and customer expectations.

5. **Seek an invested senior leader** as a champion for the pilot. They should hold the team accountable, fly cover as needed, promote the pilot, and help the team and leader to understand the importance of the project and the value of the improvements made.

6. **Designate a project leader** or coach who has good technical and social Lean skills. The team will require training, education, and coaching on principles, systems, and tools, and a credible team leader who has good experience with past Lean implementations and understands the nuances of being a change agent will be important to the project. If you do not have someone with the necessary Lean skills internally, consider bringing in an outside facilitator to co-lead the initial effort as well as developing an internal candidate.

7. **Select pilot improvement team members** who will energize and advance the project and potentially influence others outside the pilot about its value. These team members will become your "A" team players for the earliest projects. They should reflect a reasonable representation of the value stream personnel associated with the area/system/process you've selected. You might also consider including a team member who doesn't know the pilot area/system/process well to be new eyes who will ask different questions. Remember, another important goal for the pilot is to develop team members who can apply what they have learned to subsequent improvement projects; therefore, consider carefully when selecting your pilot team members.

Figure 10.1 provides some additional important points that should guide your thinking during your "inch wide and a mile deep" pilot. Remember

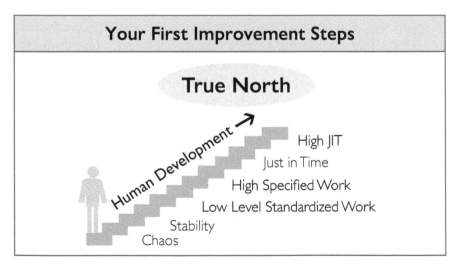

FIGURE 10.1
Basic steps in the transformation process begin with a focus on True North.

the overall pilot effort should be focused on making value flow faster in the area/system/process selected for improvement, reducing the time between paying and getting paid, if you will. The steps to do that may begin somewhere near or at the bottom and then they will continue upward. Simply stated, your team should start by understanding the current state. Then they will envision what the ideal state would look like (True North), identify any barriers to achieving the ideal state, and then conduct a series of experiments to move closer to the ideal state in the target area/system/process over time. We recommend that you develop a current and future state value stream map for your pilot, and then develop an improvement plan (an A3 document is a good method to document your improvement plan) that includes clear expectations of what will be achieved when the future state is put in place. For example: "Production lead time will go from twelve days to six days," or "Patient elapsed time for a clinic visit will go from 188 minutes to less than 65 minutes." Your improvement plan should outline the changes you intend to make and by what date, giving you a roadmap and a schedule to guide the pilot team. We will provide a bit more guidance on the purpose and use of the value stream maps in the paragraphs below.

Figure 10.1 also suggests how to think about any Lean transformation process. If your current state studies indicate that the pilot area/system/process is in a state of chaos, the first improvements your pilot will need to address must be aimed at bringing it to a stable state. For example, if machines are unreliable or numbers of staff members per day are unpredictable you will not be able to sustain improvements. If things are stable, the next check is to ensure that standardization is in place and documented standards are being universally followed.

Questions to consider when seeking to ensure that standardization is in place:

1. Do different people do the work the same regardless of the day, shift, time of month, quarter, etc.?
2. Is the output from the area/system/process consistent day to day, hour to hour, or is there variation?
3. Are the agreed upon methods for performing work documented and used regularly to verify that the standards are being met?

Standards can relate to many things, such as quality, time to perform work steps, amounts of acceptable inventory to maintain, or cost to deliver the product or service. Remember that without stable, standardized systems and processes, you have *no* baseline for continuous improvement. Your experiments will be difficult to conduct or prove out if things are shifting like beach sand in your pilot environment. Some early tools aimed at developing stability and standardization include 5S, total productive maintenance, standardized work, training within industry (TWI), standardization of changeovers and set-ups using SMED methods, and poka-yoke for mistake-proofing. This is not an all-inclusive list, but it should give you a sense of some of the tools you will need to familiarize yourself with as you proceed.

Once stable and standardized, the focus moves toward continuous flow at the pull of the customer. Remember, the ideal is flowing product or service JIT, meaning one by one, at the pull and pace of the customer, with no waste included. The key is to keep systematically working toward that ideal. Your current-state map will help you to see how far away from that ideal you currently are. If you have conducted your upfront value stream analysis correctly, you should have discovered whether the existing flow of product or service is based on push or pull (or a mix of the two) and you should have developed an understanding of takt time (the customer's expected pace of product or service delivery). You will have also calculated the current-state elapsed lead time as well as the portion of lead time that is value and the portion that is waste. The current-state map and associated data should give ideas on where your greatest barriers to ideal, waste-free flow are and should stimulate discussion on what the most logical next steps are to move the pilot effort a step (or steps) toward ideal. Once agreed upon, these next steps should be reflected in your future-state map and captured in more detail in your improvement plan/schedule for the pilot area/system/process. We recommend that each iteration of current state to future state take no more than six to nine months, so you benefit from the improvements as soon as possible and gain new learning that can inform your next round of improvement.

Tools and methods that are associated with developing or enabling continuous flow and pull include, but are not limited to:

1. Cellular production/service layouts
2. Kanban systems when true continuous flow is not possible, such as for re-supply from a stockroom or from internal or external suppliers

3. Quick changeover (SMED) to drive process flexibility and ability to cut batch sizes in pursuit of one-piece flow
4. Heijunka (levelized scheduling)
5. One-piece flow of information
6. Cross-training to enable flexibility of staff to perform all tasks and/or share work in a variety of ways to meet takt time
7. Andon systems to indicate when flow is disrupted/trigger a help chain
8. Visual management methods to track ahead or behind status and trigger problem solving
9. Mistake-proof devices to promote quality at the source
10. Daily and/or gemba management systems to establish a cadence and accountability for checking the flow and ensuring desired flow and adherence to standards, and to verify problems are being discovered, documented, and addressed in a timely manner

Base Your Pilot on the Five Principles

It is worth taking a minute to relate your "inch wide, mile deep" pilot efforts back to the five principles in the Continuous Improvement dimension of the *Shingo Model*. An appropriate goal for your pilot is to embody all five of these principles in your pilot area/system/process.

In your initial study before the pilot, as well as during your implementation phase, you need a laser-sharp Focus on Process to understand the details of how you deliver value, where you are on the continuum of improvement, what your gaps are, and what effects your changes are having. Your efforts to go to where the work is done, to study, analyze, and gain a first hand knowledge of how value is created and where waste occurs and interrupts the flow of value is exactly the kind of efforts this principle is based on.

Directionally your pilot changes should move you in the direction of Seek Perfection (True North) and the move from current state to future state should be conducted as a series of experiments that reflect the principle Embrace Scientific Thinking. You will also want to build into your pilot management oversight and problem-solving systems that are founded on scientific thinking, encouraging the rapid identification of value interruptions and supporting processes for quickly enacting structured problem solving to address these value inhibitors.

Your pilot efforts should also have carefully considered how to develop Quality at the Source thinking within your revised systems or processes.

You may ask questions such as: Within the pilot, is it extremely clear what constitutes acceptable quality at each step? Have I enabled people with the knowledge and tools to make a definitive call about the quality of work they do before it moves forward? Have I empowered team members to stop the line when problems occur so nothing escapes to a downstream area and issues can be reviewed and analyzed while they are still fresh?

Finally, as mentioned earlier, your pilot should be moving you stepwise in the direction of Improving Flow & Pull. You may not develop end-to-end continuous flow based on pull of the customer in your initial pilot, but directionally the changes you make should be moving you or preparing you to move closer to this state.

HOW WILL YOU MEASURE AND PROMOTE?

We have mentioned that the pilot effort should be considered a structured experiment in pursuit of an intended outcome. This implies the need to regularly measure progress against the pre-set goal(s) for the pilot experiment. For example, if the goal of the pilot is to reduce the start-to-end time for patient visits by 50 percent, you will need to regularly measure this time throughout the pilot to track and understand progress toward achievement of your goal. As you conduct various kaizen activities in support of your implementation plan, your measurement also provides an important way to help people understand cause and effect. For example, "When we implemented the idea to conduct check-in at the triage station our measure thereafter shows we reduced the lead time by 15 percent."

A word of caution when measuring improvement: Be sure to think upfront about what and how you will measure. If you choose a metric but then discover it will be too difficult or disruptive to perform the measure or gather the data needed for the measure, this is problematic. Similarly, if you choose a measure that others believe is not important to the experiment, is too difficult to interpret, or is controversial, the pilot efforts may be questioned.

Appropriate promotion of your pilot effort is also critical. Remember, although your pilot may be small and focused, many eyes will be on it and you need to think of it as a key step to pave the way for great acceptance of future improvement. Therefore, you will need to consciously think about how you will communicate before, during, and after the pilot takes place. A 5W1H approach may help you to think this through:

1. *Who* needs to know *what*?
2. *When* do they need to know?
3. *Where* will the communication have the most impact?
4. *Why* should they be interested?
5. *How* should we approach the communication of our efforts?

You may need different promotional and communications materials and methods depending on your stakeholder audiences. It is worth thinking this through and asking some of these questions early on in your improvement journey. Finally, be sure to engage with people to validate what you are communicating and promoting. Don't assume everyone understands the need to change the same way you do. Remember, we learn in different ways and at different rates. As a leader, keep in mind the phrase "perception is reality" as you prepare your organization to adopt the technical and social changes that come with the *Shingo Model*.

Below are a few examples of how "perception is reality" might play out if you are not aware and do not manage it.

1. Perhaps your senior executive team is not in close touch with what is really happening day-to-day in your value-delivering systems and processes. This could mean that they will be vulnerable to arguments against implementing Lean brought to them by naysayers. How will you address this?
2. Middle managers, who are typically the owners of your various systems and processes, may feel threatened when changes are suggested for these systems and processes. They may also feel they are losing power or control when you start to engage Lean teams in suggesting improvement in areas/systems/processes they feel they own. Again, how will you head this off?
3. Some areas may view a principle like Quality at the Source as threatening their turf. Does it mean, for example, that when people are trained and empowered to check their own quality the Quality Inspection function will no longer be needed? How will you handle this type of fear?

HOW WILL YOU DEVELOP INTERNAL CAPABILITY?

Key to the *Shingo Model* is the idea of continuously developing and improving the capabilities of people. Ultimately, the goal is to establish a robust learning organization, one with a population of problem-solvers, a vast stable of people who are practiced and confident in their ability to see and solve problems, and who resourcefully and creatively drive the organization closer and closer to perfection over time. This does not happen by chance, and you will need to think deeply about how you will develop this level of internal capability.

Foremost, your goal should be to avoid creating a culture where it is normal to have continuous improvement done *to* your employees or *for* your employees. Instead, a healthy improvement culture is anchored in changes done *by* your employees. This tacit learning is how people gain the skills, knowledge, and confidence to fuel the improvement engine going forward.

At the outset of your improvement journey, engaging people can feel challenging as there may be limited internal knowledge of and experience with Lean principles, systems, tools, and culture. You may need to leverage some outside support for initial training and/or to facilitate initial improvement efforts. The caveat here is to be certain this outside support is actively developing your internal staff to become the next Lean leaders and teachers so that the organization can become self-sufficient in its improvement journey as soon as possible.

You may also choose to add one or more Lean roles into your organization at the outset or along the way in order to train, coach, and facilitate early efforts. Again, choose carefully and be certain that this role is intended to help and facilitate improvement activities, not to *do* them. When a continuous improvement role is added to an organization, we often find that people believe improvement is only to be done by the improvement person/department.

There are many additional ways you can promote and encourage Lean learning and practice for members of your organization. We will share a few here, but there are many more. We strongly encourage you to create opportunities for people at all levels in your organization to engage with others in the Lean community. These practitioners are a great source of insight, knowledge, and practical experience when it comes to the realities of developing an improvement culture founded on proven principles.

Ideas to Promote CI Learning in Your Organization

1. Book study with a team or instructor-led classroom activities are a great way to provide people with the basic thinking and an introduction to tools and principles. However, these methods cannot create deep learning on their own. You will need to couple this type of learning with constant hands-on practice.
2. Visit customers or suppliers to understand your extended value stream. This will help team members gain a broader view of the impact of their work and can also provide opportunities to see how the key stakeholders of the business are addressing improvement efforts.
3. Attend Lean conferences, workshops, or webinars to engage with other Lean practitioners. Shingo training workshops are offered frequently throughout the year, virtually and on-site, and can also be hosted as private classes at any organization.
4. Local Lean consortia are another avenue to develop your capabilities. Through membership in a Lean consortium, your organization will participate with others on the journey, accessing additional resources and opportunities for learning.
5. Seek out a mentoring relationship with an external colleague or consultant. Mentoring is a good way to develop individuals and help them see beyond conceptual blind spots. Experienced eyes can guide the inexperienced and the relationship can be customized to meet the needs of the mentee.

HOW WILL YOU EXPAND THE IMPROVEMENT BOUNDARY?

Remember that one of the intents of an in-depth pilot is to pave the way for broader and deeper continuous improvement efforts after the initial pilot is established. This early success helps deepen the understanding and acceptance of both the technical and social aspects of Lean. It also provides a safe environment for people to gain practical experience that they can then use in the organization's future improvement efforts.

You will want to leverage the knowledge and experience of these early practitioners to expand your improvement bandwidth. This will allow them to assist and/or lead the next teams and improvement projects and to act as teachers and coaches for newer participants. When you consciously spread improvement in this way, you are building your base of capable problem solvers and ensuring that your ongoing Lean projects are founded on a consistent understanding of principles, systems, tools, and desired behaviors.

As you expand your improvement boundaries, it is also important to review each new targeted effort through the lenses outlined earlier in this chapter. That is, understanding upfront your new current state and gaps, checking that the proposed project will help address your most important needs, and verifying that the work will be conducted and measured in accordance with scientific principles and that the efforts will provide further opportunities for organizational learning and human development. The ongoing involvement of executives and managers and a commitment of resources will also be necessary to effectively continue your improvement efforts.

Ultimately, the boundaries of your improvement efforts should extend outside your four walls, linking to customers on one end of your extended value stream and to suppliers on the other end. A word to the wise: before setting expectations for joint continuous improvement with suppliers or customers it is necessary to get your own house in order! When your systems and processes are stable, standardized, and starting to reflect appropriate use of Lean principles and methods, only then will you possess enough knowledge and credibility to invite your customers and suppliers to the game.

SUMMARY

During a panel discussion on innovation a few years back, Dan Ryan, then VP of Corporate Operations at Raytheon, made a striking statement. He said, "Innovation equals continuous improvement. Our people are the source of our innovation." His point was not that technology is unimportant, but rather that it is engaged team members who are the creative force behind Raytheon advancements. This was a powerful message coming from a top executive of one of the world's most highly innovative technology companies and 2008 recipient of the Shingo Silver Medallion.

Kevin Spradlin, assembly leader at Whirlpool, made a similar statement as a guest on the *Old Lean Dude* blog. Talking about a successful effort to reduce the cost of a Whirlpool dishwasher line under the coaching and guidance of Mr. Oba, who was the former director of TSSC, Spradlin said,

> Whirlpool's Findlay Operations are on a journey to become a continuous improvement organization. We want to be the undisputed choice when someone goes to buy a dishwasher, and they only consider buying our product. We believe in putting the customer first, the dedication of our people and staying committed to continuous improvement wherever the work is done.[5]

As you proceed on your Shingo-based improvement journey, consider the messages of Dan Ryan and Kevin Spradlin. The technical science embodied in the continuous improvement dimension of the *Shingo Model* can only succeed when you develop a strong appreciation for the five principles that make up the dimension *and* when you are mindful of their interdependency with the principles captured in the other two dimensions. Building a culture that reflects people proposing and making changes in the right direction, in a sustainable way, enabled by systems and tools that foster value creation, ideal behaviors, organizational learning, and human development doesn't happen by chance. It requires a commitment to the long term as well as repeated cycles of experimentation that focus on increasing value by making work easier, better, faster, and less expensive for those who perform the work.

---

[5] Kevin Spradlin, "Always Made in America," *Old Lean Dude* (blog), April 2014, https://oldleandude.com/tag/continuous/.

Glossary

3M   Three conditions for improvement in Japanese: *muda*—"waste," as in the seven wastes (see later), *mura*—"uneven" or "inconsistent," and *muri*—"impossible" or "unreasonable."

3P   New product development process: product, process, and production. Also referred to as new production preparation. The objective of 3P is to design products that meet the requirements customers want at an acceptable price, and that can be manufactured with simple, low-cost processes at the same rate as sales demand requires (i.e., manufactured to takt time).

4 Elements of Process   Shingo's four elements of process are: 1) processing: physical change in the material or its quality (assembly or disassembly), 2) inspection: comparison with an established standard, 3) transportation: movement of material or products (change in location), and 4) delay: period of time during which no processing, inspection, or transport occur. Within delay are two subsections: 1) Process delay: an entire lot waits while the previous lot is processed, inspected, or moved, and 2) lot delay: in lot operations, while one piece is processed, the others wait to be processed or for the rest of the lot to be done. This phenomenon occurs in inspection and transport as well.

4M   The four elements of production: man (person), method, material, and machine.

5S   Workplace organization and standardization are fundamental to any manufacturer. The five pillars of 5S are Japanese words that describe a progressive process for creating and sustaining a stable, repeatable production (or administrative) environment. They are:
- *Seiri:* to sort out unnecessary, seldom used, or excess items
- *Seiton:* to set locations for all needed items, placed close at hand according to use
- *Seiso:* to scrub, shine, and sweep. Keep an area free of clutter and foreign material
- *Seiketsu:* to standardize methods for locating and maintaining order to create a habit of 5S

- *Shitsuke:* to sustain the 5S process through promotion and management walk-throughs

5W1H The basic questions to ask when describing a condition or problem: who, what, when, where, why, and how.

5 Whys The most basic form of root-cause analysis involves asking "why" five times to get to the root of a problem. For example:

1. Why was the order late? *Because it was missing a part.*
2. Why was the part missing? *Because the machine was broken.*
3. Why was the machine broken? *Because the brushes on motor were bad.*
4. Why were the brushes bad? *Because they had not been replaced.*
5. Why had they not been replaced? *Because they are not on our preventative maintenance schedule.*

Six Sigma A highly disciplined use of quality control tools employed to improve and reduce defects to less than 3.4 per million or better.

6-Step Problem Solving A basic model for defining and solving problems: 1) define the problem, 2) analyze, 3) brainstorm potential solutions, 4) select a solution and create a plan for implementation, 5) implement the solution, and 6) evaluate the solution.

7 Wastes Seven undesirable conditions, or *muda*, that can arise in any process: 1) storage or inventory, *2)* transportation, 3) overproduction, 4) processing, either inadequate or unnecessary, 5) motion, 6) defects (correction), and 7) waiting. With careful observation and appropriate countermeasures, these wastes can be greatly reduced or eliminated to improve quality and productivity. For a more detailed description of each of the seven wastes, refer to each term in the following.

8th Waste Loss of human creativity. Although not a waste in an industrial engineering sense, the eighth waste was added when western companies started adopting practices following the Toyota Production System. While Shigeo Shingo never called the loss of human creativity the eighth waste, he did teach that most traditional organizations ignore the thinking and creativity of their workers.

A

A3 or A3 Report A visual manifestation of a problem-solving thought process represented on a single sheet of A3 size paper (an A3-sized paper is an international standard close in size to an 11 × 17-inch sheet). It involves scientific thinking and continual dialogue between the owner of an issue and others in an organization. The single sheet captures the problem, analysis, corrective actions, and action plan, and is often used with graphics. Different versions of A3 reports have evolved into a standard method for summarizing problem-solving exercises, status reports, and planning exercises like value stream mapping.

Administrative Kaizen The application of continuous improvement to non-production functions, such as accounting or sales, using best practices of lean to reduce waste in administrative areas.

Andon Visual or audible signal that indicates changing conditions within a factory, detection of a defect, such as set-up, delays, stock-outs, need for assistance, deviation from schedule, or need for overtime. An Andon board often refers to a board that is always visible to the people in a work area, which displays the status of work throughout the working time.

AQL Acronym that stands for *acceptable quality level*. A traditional supply-driven forgiveness for part non-conformance, (typically 1–2 percent) that was once considered acceptable. AQL is now an oxymoron in the Lean lexicon.

Autonomation Referring to retrofitting machines with self-checks for defects, enabling workers to perform more useful tasks than waiting for machine faults. Also known as *automation with a human touch*, or Toyota-style *jidōka*.

B

Behavior The way in which one acts or conducts oneself, especially toward others. Behavior can be observed, described, and recorded.

Behavioral Benchmark Large groups of behaviors used to break down principles into manageable groups. They can be further broken down into core ideal behaviors, the systems that drive them, and key behavioral indicators (KBIs), or what we measure.

Baka-Yoke Japanese term that translates to "fool-proofing," the original concept of mistake-proofing. To respect the intelligence of workers, Shingo changed the term to *poka-yoke*, or "mistake-free."

Batch A process that requires more than one piece at a time to be processed (e.g., ovens, mixers, automatic machining processes often require large batches to be processed at once). Batches consequently create process delay before a subsequent operation.

Batch Delay   Delay caused by material waiting to either side of a process when a batch is produced. This is also called *lot delay*.

Blitz Kaizen   A structured, event-driven improvement process intended to produce rapid lean improvements. The main advantage of this method is that a focused project concentrated within three-to-five days can produce significant gains. The disadvantages are that the improved process must typically be shut down during the improvement, and that productivity gains may not be sustained without a concerted follow-up.

C

Catch-Ball   Sharing information in both directions, often in the process of Hoshin planning/policy deployment. The idea is to pass the initial information (usually goals and plans) to the various levels of the organization, where they will "catch" it and review, ask questions, and supply comments and feedback, before "throwing" that input back through the levels who will "catch" what is passed back. This can be an iterative process and is aimed at helping to involve the right people in developing plans and ensuring common alignment to plans.

CEDAC   Acronym for *cause-and-effect diagram adding cards*. An improvement method based on the fishbone diagram but enhanced through a systematic evaluation process using color-coded cards.

Cells   The layout of machines/work steps in a tight sequence, usually in a U-shape, to make it easier to achieve one-piece flow and to allow for flexible deployment of people to staff the cell.

Cellular Manufacturing   Production organized by sequence of operation, with all required machines to produce a particular family of products grouped together in a U-shaped arrangement. This concept can also be used in support settings.

Changeover   Activities required to change a process or machine between different products or process steps. Long changeover time is typically associated with long production runs, batching, and inventory stagnation.

Coaching Kata   The repeating routine by which Lean leaders and managers teach the improvement steps to everyone in the organization. The teacher or coach gives the learner procedural guidance, not solutions, that help the learner to be successful in overcoming obstacles.

Continuous Flow   Production organized by operation sequence and balanced to create a level part flow with minimum lot and process delay within the boundaries of the process. Perfect continuous flow is based on one piece at a time processing with no inventory accumulating between processing steps and a flow rate that reflects the demand rate (takt time).

Continuous Improvement   A philosophy and system for improvement that assumes any current condition can be improved, and that given a sound understanding of system and process wastes, all team members will be motivated to make ongoing improvements that benefit both them and their customers. The ideal continuous improvement involves all of the people, in all parts of the organization, making improvements all of the time.

Culture   The cumulative behaviors within an organization, and the purpose, systems, values, and beliefs that drive those behaviors.

Cycle Time   The length of time to complete an operation (as in *elemental* cycle time) or the sum of elemental cycle times for a single part or product. Cycle time should not be confused with *takt* time, or the rate of demand.

D

Defects   One of the seven wastes of the Toyota Production System, manifested variously in scrap, rework, and additional inspection. Also referred to as the waste of correction. See also *Isolated Defects* and *Serial Defects*.

Detect Device   An objective inspection device that detects the presence of a defect after it has been made but before it is passed on to the next operation.

E

Empowerment   A critical aspect of continuous improvement that involves three essential steps:

1. Training team members to identify waste and use appropriate countermeasures to reduce waste.
2. Creating a favorable environment for practicing continuous improvement, including policy deployment, supporting measures, and budgeted resources for improvement.
3. Keeping all team members involved in improvement through leadership and appropriate recognition and rewards.

EOQ   Acronym for *economic order quantity.* A traditional way of opti-mizing inventory levels by balancing changeover and inventory carrying costs. Lean thinking asserts that it cannot be truly eco-nomical to produce inventory that must be stored, and that an EOQ of one must be the ideal condition for every process.

Everybody, Every Day   The basic premise of improvement that assumes the power of Lean is derived from many improvements, both small and large, accumulated over time by all team members.

External Set-Up   Set-up (or changeover) tasks that may be accomplished while the process is still running. Moving internal set-up tasks to external can greatly reduce the time that the process is stopped.

F

FIFO   Acronym for *first in, first out.* A key ordering priority for produc-tion that reflects the stability and repeatability of a process.

Flow   Term first used by Henry Ford to describe a production system in which material is processed continuously, without storage or delay, from its initial operation to final production and shipment.

FMEA   Acronym for *failure mode and effect analysis.* FMEA is an analyt-ical tool used to predict and eliminate any potential design defect in a new product in advance by analyzing the effects of failure modes of component parts on final product performance. See also *PFMEA.*

Frequent Withdrawal   A key practice of Lean manufacturing in which each subsequent process withdraws needed material on a frequent basis creating a clear production instruction to each producing department.

Fundamental Truth   Beneath every principle there is a fundamental truth that provides the basis for why a principle is universal and timeless. This truth is often the part that is the most self-evident, even though at first the fundamental truth may be difficult to identify.

G

Gemba   Japanese word meaning "real place," referring to where the actual work is done in any area of any organization.

Go See   A key philosophical guidepost of the Toyota Production System that requires direct observation as the first step to understanding a problem or current condition. In the Shingo Institute workshops, the term "Go & Observe" is utilized to emphasize the need that the activity should not be casual *seeing*, but intense *observation*.

Guiding Principles The ten Shingo Guiding Principles are the basis for building a lasting culture and achieving organizational excellence. They are divided into three dimensions: cultural enablers, continuous improvement, and enterprise alignment.

H

Heijunka Levelization of the production schedule achieved by smoothing both the volume and mix of products through a production line. Leveling the type and quantity of production over a fixed period of time enables production to efficiently meet customer demands while avoiding batching and associated wastes.

Hoshin Planning This policy deployment method focuses the organization on a few vital breakthrough improvements. The objectives and means to achieve the objectives are cascaded through the organization using a series of linked matrixes. Also known as *policy deployment* and *Hoshin Kanri*.

I

Ideal Behavior Actions that create outcomes that produce results that are both excellent and sustainable.

Ideal Results Outcomes that are aligned, that are both excellent and sustainable, and that demonstrate improvement over time.

Improvement Kata A repeating four-step routine to improve and adapt. It uses the scientific problem-solving method of plan, do, check, act (PDCA) to encourage improvement as a daily habit. The four steps are 1) determine a vision or direction, 2) grasp the current condition, 3) define the next target condition, and 4) move toward the target through quick, iterative PDCA cycles to uncover and remove obstacles. (Note: The plan is defined by the first three steps.) See also *Kata*.

Inadequate Process One of the seven wastes. Inadequate process describes any undesirable condition arising from processing itself, such as over- or underprocessing, or a process that is not needed. Also known as *unnecessary processing* or *processing*.

Inspection The act of examining or viewing, especially carefully or critically.

Internal Set-Up   Set-up tasks and associated time that must be completed while the process is stopped.

Isolated Defects   Defects that occur only once. See also *Defects*.

J

Jidōka   In Japanese, the word for *automation*.[1] Toyota developed its own version of jidōka,[2] which became one of the two pillars of the Toyota Production System. The Toyota version of jidōka is sometimes translated as *autonomation* or *automation with a human touch*. The words are pronounced identically, and the Chinese characters used are identical except in the second character of the Toyota version, there is an additional element. This additional element means *a person*, hence *automation with a human touch*.[3]

   There are three additional steps to this Toyota-style automation:
1. The machine detects defects;
2. The machine automatically shuts down the operation when a defect is detected;
3. The machine automatically notifies the operator of the problem.

   See also *Autonomation*.

Judgment Inspection   Sorting good products that meets specifications from defective products to prevent shipping products with defects to customers.

Just-in-Time   One of the two pillars of the Toyota Production System, just-in-time (JIT) combines a series of techniques designed to make and convey only what the customer needs when the customer needs it in the quantity the customer needs. Use of continuous improvement principles and tools are aimed at enabling JIT to take place at the lowest cost with the least on-hand inventory.

K

Kaikaku   A radical improvement activity, also referred to as *breakthrough kaizen*.

---

[1] 自動化.

[2] 自働化.

[3] See Shingo Shigeo, *A Study of the Toyota Production System from an Industrial Engineering Viewpoint* (Portland, OR: Productivity Press, 1989), 161–164 for an explanation of "Autonomation" and "Automation with a human touch."

Kaizen Japanese word meaning "improvement." Kaizen is the continuous pursuit by all team members for easier, better, faster, cheaper. Focus is on creating more value with less waste (*muda*).

Kanban Japanese word meaning "sign." Kanban is the basis for triggering withdrawal and production of material or work. Two types of kanban cards are used to manage production according to customers' needs:

1. *Production kanban cards* are placed with small set quantities of material (also referred to as *kanbans*), produced in advance of demand in order to provide an apparent lead time of zero to the consumer. When material is withdrawn by the consumer (downstream process), the production kanban is passed back to the producer to trigger additional production.

2. *Withdrawal kanban cards* are authorizations to withdraw a set quantity (or kanban) of material from the producer's store. These kanbans stay with the withdrawn material until that material is consumed, at which point the cards can be used to authorize an additional withdrawal from the producer. The pull system reduces the waste of overproduction by synchronizing all production processes to actual customer need.

Kata Kata has its origins in the Japanese martial arts, but it also refers to any basic form, routine, or pattern of behavior. In continuous improvement, identifying patterns of behavior and clear expectations make it easy to recognize abnormalities (problems). It also provides a basis of improvement. In Lean management, kata refers to two linked behaviors: improvement kata and coaching kata. See also *Improvement Kata* and *Coaching Kata*.

KBI An acronym for *key behavior indicators*. These are measurements that track identified ideal behaviors. See *Ideal Behavior*.

KPI An acronym for *key performance indicators*. See *Results*.

L

Lead Time The total time a customer must wait to receive a product or service after placing an order.

Lean A philosophical and methodological approach to manufacturing that strives to provide ever-increasing value to the customer through total employee involvement in the reduction of non-value-added wastes and their associated costs. Womack, Jones, and Roos first adopted the term Lean to describe manufacturing systems that are based on the principles employed in the Toyota Production System. Quoting them, "Lean . . . is 'lean' because it uses less of everything compared with mass production—half the human effort in the factory, half the manufacturing space, half the investment in tools, half the engineering hours to develop a new product in half the time. Also, it requires keeping far less than half the inventory on site, results in many fewer defects, and produces a greater and ever-growing variety of products."[4] We use Lean in the meaning that was intended when it was first coined.

Levelization   A technique that strives to create a production sequence that is leveled by both quantity and model mix. The ideally leveled factory will level production to a flow of one-piece at a time, with no like models being produced back-to-back.

Lot Delay   Delay caused by material waiting at either side of a process when a batch is produced. See also *Process Delay*.

M

Material and Information Flow Diagramming   This process provides a graphic, visual depiction of any production/work process. The goal of M&I diagramming is to highlight wastes in the current condition in order to help create a target for improvement. See also *Value Stream Mapping*.

Mistake-Proofing   Also referred to as *poka-yoke*, or "without mistake." This process recognizes the fundamental idea that everyone occasionally makes mistakes, but that these mistakes can be detected and trapped before they create a defect. The method is important not only to reducing human-related defects to zero but also to increasing the velocity of production by eliminating continuous checking and second-guessing.

Motion   Another of the seven wastes of TPS, this waste arises from any operator motion that does not move the production process forward. Examples include unnecessary walking, searching, bending, reaching, stretching, climbing, backtracking, and awkward movements.

---

[4] James P. Womack et al., *The Machine That Changed the World*, 14. Note: To our knowledge, Toyota has never used the term Lean but always uses the term *kaizen*, which means "improve," and implies that the sought improvement is constant or continuous.

MRP Acronym for *material requirements planning.* A computerized system that uses data and system modifiers to determine the quantity and timing for producing and/or ordering materials.

Muda Japanese word for "waste." See also *7 Wastes.*

Mura Japanese word for "irregularity" or "variability."

Muri Japanese word for "impossibility" and "difficulty." *Muri* can relate to both people or machines.

N

New Production Preparation This process applies the methodology of standardized work to plan and model new production according to anticipated customer requirements. See also *3P.*

Non-Value Added A generic expression used to describe any activity or cost that is not fundamental to the customer's need and that the customer would not want to pay for.

O

Objective Inspection Inspection made by someone other than the operator who performed the work.

OEE Acronym for *overall equipment effectiveness.* This key measure from Total Productive Maintenance (TPM) is a very stringent compound measure derived from three factors:

1. Machine Availability = (planned operating time—downtime)/ planned operating time
2. Performance = (ideal cycle time x pieces produced)/planned operating time
3. Quality = (planned quantity—defects) planned quantity
   OEE = machine availability × performance × quality.

An OEE of 85 percent or better will typically be considered world class. Most factories beginning a TPM process will discover initial OEEs closer to 30 percent. The most accurate measurement of OEE includes changeover time as part of downtime.

Ohno, Taiichi Taiichi Ohno is considered the father of the Toyota Production System, which is now often referred to as Lean manufacturing. Mr. Ohno studied Henry Ford's flow manufacturing system carefully and then made a critical improvement to allow for fast and easy line changes and pull systems.

One-Piece Flow   The ideal condition for any production/work system is to produce exactly what the customer needs when he needs it and with perfect quality. One-piece flow (OPF) manifests that ideal condition by achieving one-by-one part and information flow within a portion (or possibly all) of the production process. To be implemented effectively, delays from changeover and transportation must be nearly zero and defects must be zero.

Operation   An activity or specific step performed within the process to deliver a product or service.

Overproduction   The worst of the seven wastes of TPS because any part produced before it is needed will subsequently incur some or all of the other seven wastes.

P

PDCA   Acronym for *plan-do-check-act*. PDCA are the basic steps in conducting problem-solving and performing kaizen.

PFMEA   Acronym for *process failure mode and effects analysis*. PFMEA is a methodical approach used for identifying risks on process changes. See also *FMEA*.

Physical Inspection   Inspection by the means of measuring devices.

Poka-Yoke   Japanese word meaning "mistake free." Also known as mistake-proofing, this method for trapping human mistakes before they become defects has the effect of reducing these types of defects to zero. It is a key method in support of the principle Quality at the Source.

Policy Deployment   Any of several means for directing improvement through a systematic cascading of goals to specific achievable projects to be completed within a set time period, usually one year. This process is key to creating organizational alignment. See also *Hoshin Planning*.

Prevent Device   Any mistake-proof device that is able to trap a mistake before it can create a defect.

Principle   A foundational rule. Principles are universal and timeless, evident, and govern consequences. They help us to see the positive and negative consequences of our behaviors and allow us to make more informed decisions.

Process   The complete series of operations required to deliver a product or service. This could be to complete a design, produce a product, treat a patient, etc.

Process Delay Process delay is created when a batch of material must be completely processed at one operation before being passed on to a subsequent operation. While workers and machines may be very busy, the velocity of the inventory (and hence cash flow) is very sluggish. See also *Lot Delay.*

Process-External Inspection Inspection carried out at a different process from where the work was performed. Also called a *judgment inspection.*

Process-Internal Inspection Inspection carried out at the same process where the work was performed. Also called a *source inspection.*

Process/Operation Grid Created by Shigeo Shingo to clarify the two-dimensional nature of any process. He asserted that a process was much more than the sum of operational steps and times. It also included significant delays that were incurred from each of the seven wastes. Up to 95 percent of process time, Shingo showed, was a delay caused by one or more of the seven wastes. Therefore, the proper focus for improvement was on the 95 percent waste rather than the 5 percent operation time.

Pull System A system that operates by receiving signals when more production is needed. Taiichi Ohno, creator of the Toyota Production System, studied the ordering pattern of American supermarkets to create the pull system. His belief that material replenishment be based entirely on customer withdrawal was a major improvement in managing production and inventories.

Push System The traditional method for scheduling production involves order launching and expediting beginning at the lowest level part and then "pushing" material through intermediate work steps and/or stockrooms until it is finally needed for a customer order. This method creates huge excess inventories and drains productivity from factory workers and machines.

Q

QFD Acronym for *quality function deployment.* QFD is a structured, matrix-driven tool to identify the key attributes and features to be provided in a product/service, and then cascade these priorities through the product/service development process.

Queue Time The amount of time a product waits before beginning the next step in a process.

Quick Change   Any of a number of practices that reduce the changeover time of a process by improving the preparation, tools/equipment/ supplies setting, and adjustment required for set-up.

R

Results   A measurable outcome, either successful or unsuccessful, that results from the implementation of tools and systems. See *KPI*.

Rework   This is one of the results of the waste of defects, creating a huge, often hidden cost in both time-to-market and margin. In addition to tracked rework, there is typically a large increment of operator accommodation that is not tracked.

S

Sensory Inspection   Inspection performed by means of the human senses.

Serial Defects   Defects that occur repeatedly. See also *Defects*.

Set-Up Reduction   Reducing the amount of time that a machine or process is down during changeover. Set-up time is defined as the time to change from the last good piece of the previous order to the first good item of the next order. With concerted effort, set-up times can be reduced by 95 percent or more, enabling machines to economically process very small quantities between changeovers and creating greater flexibility. Also referred to as *changeover reduction*.

Shingo, Shigeo   Japanese industrial engineer, co-creator of TPS, and Taiichi Ohno's shop floor improvement person at Toyota. Shingo wrote extensively on TPS and is credited with developing the concepts single-minute exchange of dies (SMED or quick changeover) and poka-yoke. In 1988, a manufacturing prize, the North American Shingo Prizes for Excellence in Manufacturing, now, simply, the Shingo Prize, was created in his honor by the Shingo Institute.

Single-Piece Flow   Method where a product or service is moved through operational steps one at a time, without interruptions, backflows, or queue time. Also referred to as *one-piece flow*.

SMED   Acronym for *single-minute exchange of dies*. A method for rapid changeover pioneered at Toyota that enabled them to reduce changeover time from hours to minutes. The principles of this method apply to any machine or process, and can, in the words of creator Shigeo Shingo, "reduce changeover time by 59/60ths."

Source Inspection   An inspection performed at the place of work when the work is being done. The purpose is to detect and prevent the source of an error or defect. Source inspections act to mitigate or eliminate the causes of errors and defects. If you can contain the errors that cause defects, you can radically reduce or eliminate defects for many future production cycles.

Stagnation   Term used to describe any information or material that is in a delayed state. Traditional manufacturing uses the phrase *work in process*, but if material or information is not actually being processed, Lean thinking views it as stagnant.

Standard WIP   Acronym for *work in progress*. Standard WIP is the minimum amount of inventory needed to avoid delays between operations. Using standardized work, this buffer may be calculated and maintained. Status of Standard WIP also quickly signals whether the process is operating as designed.

Standardized Work   One of the key components of just-in-time, this method carefully organizes work sequence and standard WIP to match the expected rate of customer consumption (takt time). Focus on motion and machine improvement are critical to providing best productivity and quality and to making the job easier and safer.

Stop the Line   Defects must be corrected immediately in a Lean system. This requires that any operator encountering a defect be required to stop production in order to correct the problem before it or any subsequent defects are passed on to the next process.

Storage   One of the seven wastes, storage involves not only formal stockrooms, but all places that material will be staged, picked up, and put down, including bench tops, carts, pallets, inspection stations, etc. Stored inventory is viewed as stagnation by Lean manufacturers.

System   A collection of tools working together to accomplish an intended outcome.

T

Tacit Learning   Lean manufacturing is learned by doing. While the seven wastes are universal, no two factories or organizations are alike. Every team member must understand waste according to their specific job and must personally experience the benefits of eliminating waste through the use of Lean countermeasures.

Takt Time   Takt time is the time given by the customer to produce a particular product or service. It represents the drum beat to which the producer of the product or service should be calibrated and it may change from month to month. Takt time is calculated by dividing the minutes (or seconds) of available work time by the daily customer requirement. For example, a workday of eight hours, less breaks and clean-up, is 450 minutes. If a factory must produce forty-five products per day, then takt time is ten minutes.

Tool   A single device or point solution that accomplishes a specific task.

TPM   Acronym for *total productive maintenance*. This is a structured method for reducing losses to productivity and quality resulting from poor equipment maintenance. For capital intensive factories, this method can generate very large gains. Good 5S is often viewed as an entry step for TPM.

TPS   Acronym for *Toyota Production System*. TPS was pioneered at Toyota Motor Works over the past fifty years. This system provides the philosophical, management, and technical basis for Lean.

Transportation   One of the seven wastes, transportation waste results from excessive movement of material or information, particularly intercompany movement. Intercompany travel distances of miles are not unusual and can often be slashed to feet through the use of continuous flow and cellular manufacturing.

True North   A term used to indicate the right direction for improvement. This direction should be based upon providing perfect product just-in-time according to the customer's desired product or service. It should be visual and self-managing and continuously searching for higher quality, lower cost, shorter lead times, and greater selection for the customer.

U

Unnecessary Process   This waste refers to any operation that is unnecessary, redundant, or done in excess. Also referred to as *inadequate process*.

U-Shaped Cell   Wrapping a production line around itself into a U. This design enables workers to share variable cycle times and work volume more effectively, and to enable pick-up and delivery of material to occur without stopping production.

V

Value-Add   Term used to describe any activity that is seen as valuable by the customer. It is also used in calculations as a lean productivity measurement describing the difference between net sales and purchases:

$$\text{value-added} = \text{net sales—purchases}$$

Value Analysis   Method that uses Lean design tools to establish the best value and selection in a product design and producibility according the market needs and price point.

Value Engineering   A design practice aimed at honing a product design to specific customer requirements, usually with the intent of reducing costs by removing unneeded features.

Value Stream Map   A graphical method for highlighting improvement opportunities to a process or system that was developed to create a consensus for improvement activities. The process endeavors to improve material and information flow in support of smooth production flow. See also *Material and Information Flow.*

VCS   Acronym for *visual control systems.* The practice of VCS creates a self-managing process where operational status is clear at a glance, often from a great distance. Standardized visual devices enable workers to work without hesitation or delay, and to communicate needs quickly. It also helps people identify abnormalities quickly. Sometimes referred to as *visual management.*

VRP   Acronym for *variety reduction process.* A design tool that strives to maximize product selection to the customer by minimizing any variety that has not been requested by the customer. VRP targets the intersection of two cost curves, value engineering and standardization, to eliminate costs created by inadvertent causes such as different design periods and designer preferences. Organizations using VRP typically reduce part and/or process count by more than half.

W

Waiting   One of the seven wastes. Waiting is perhaps the most hidden waste as it relates to material, people, or machine.

Waste   Identification of the seven wastes is the key to improvement in any production system. These are specific and measurable industrial engineering wastes that together represent an organization's improvement opportunities. See *7 Wastes*.

X

X Matrix   The X matrix is a policy deployment tool that assembles all corporate objectives, improvement priorities, measures, and financial targets on a single page.

Y

Yesterday's News   An expression used in value stream mapping that refers to computer reports that are often out of phase with actual production and material flow, rendering the information as useful as "yesterday's news."

Z

Zero   A concept that targets zero for defects, changeover, zero production lost to equipment failures, zero accidents, and zero waste due to elimination of each of the seven wastes. A hypothetical process.

Zero Quality Control   The term Shigeo Shingo gave to the source inspection and poka-yoke process (and the title of his book on the subject) to denote that the target of poka-yoke will result in zero defects, and also in zero concern for quality control—type judgment inspections.

Bibliography

Covey, Steven. *The Seven Habits of Highly Effective People*. New York, NY: Simon & Schuster, 2020.

Deming, Edwards W. *The Red Bead Experiment with Dr. W. Edwards Deming*. The Deming Institute Video, 9:00, https://deming.org/deming-red-bead-experiment/.

Hirano, Hiroyuki. *JIT Factory Revolution: A Pictorial Guide to Factory Design of the Future*. Portland, OR: Productivity, 1988.

Imai, Masaaki. *Gemba Kaizen: A Commonsense Approach to a Continuous Improvement Strategy*. New York, NY: McGraw-Hill, 1997.

Liker, Jeffrey and Michael Hoseus. *Toyota Culture: The Heart and Soul of the Toyota Way*. Columbus, OR: McGraw-Hill Education, 2008.

Merriam-Webster's Collegiate Dictionary, 11th ed. Springfield, MA: Merriam-Webster, Inc., 2003.

Moments of Truth. DVD. Directed by Bruce Hamilton. Boston, MA: GBMP, 2009.

Ohno, Taiichi. *Toyota Production System: Beyond Large-Scale Production*. Portland, OR: Productivity Press, 1988.

Plenert, Gerhard. *Discover Excellence: An Overview of the Shingo Model and Its Guiding Principles*. Boca Raton, FL: CRC Press, 2018.

Robinson, Alan G. *Modern Approaches to Manufacturing Improvement: The Shingo System*. Portland, OR: Productivity Press, 1990.

Shingo Institute. *The Shingo Model, Version 14.5*. Logan, UT: Utah State University, 2020.

Shingo, Ritsuo. *Go & Observe*. MP4 video, 2:54, from an interview by the Shingo Institute, April 2016.

Shingo, Shigeo. *Zero Quality Control: Source Inspection and the Poka-Yoke System*. Translated by Andrew P. Dillon. Cambridge, MA: Productivity, 1986.

Shingo, Shigeo. *The Sayings of Shigeo Shingo: Key Strategies for Plant Improvement*. Translated by Andrew P. Dillon. Cambridge, MA: Productivity Press, 1987.

Shingo, Shigeo. *A Revolution in Manufacturing: The SMED System*. Translated by Andrew P. Dillon. Cambridge, MA: Productivity Press, 1988.

Shingo, Shigeo. *Non-Stock Production: The Shingo System of Continuous Improvement*. Translated by Andrew P. Dillon. Cambridge, MA: Productivity Press, 1988.

Shingo, Shigeo. *A Study of the Toyota Production System: From an Industrial Engineering Viewpoint*. Translated by Andrew P. Dillon. Portland, OR: Productivity Press, 1989.

Shingo, Shigeo. *The Shingo Production Management System: Improving Process Functions*. Translated by Andrew P. Dillon. Cambridge, MA: Productivity Press, 1992.

Shook, John and Mike Rother. *Learning to See: Value Stream Mapping to Add Value and Eliminate Muda*. Boston, MA: LEI, 1999.

Spear, Steve and H. Kent Bowen. "Decoding the DNA of the Toyota Production System." *Harvard Business Review*, Sept–Oct (1999): 98.

Spradlin, Kevin. *Old Lean Dude* (blog). "Always Made in America," April 2014. https://old-leandude.com/tag/continuous/.

Taylor, Alex. "GM: Death of an American Dream." *Fortune Magazine*, November 2008.

Toast: Value Stream Mapping. DVD. Directed by Bruce Hamilton. Boston, MA: GBMP, 2009.

Womack, J. P. et al. The Machine That Changed the World: The Story of Lean Production—Toyota's Secret Weapon in the Global Car Wars That Is Now Revolutionizing World Industry. New York, NY: Simon & Schuster, Inc., 1990.

Recommended Reading

This recommended reading and watching list provides Lean reference material (as well as Lean healthcare resources) that we have found helpful along our Lean journeys. The list is not meant to be all inclusive and we encourage you to cast a wide net as you read and research Lean publications. Be sure to visit https://shingo.org/publication-award/ to connect you to publications that have been recognized by the Shingo Institute for excellence in advancing the body of knowledge regarding organizational excellence.

5S-5 Challenges: Making Lean's First Improvement Last. Directed by Bruce Hamilton. DVD. Boston, MA: GBMP, 2009.
    Understanding sustaining workplace organization

Bahri, Sami. *Follow the Learner: The Role of a Leader in Creating a Lean Culture*, 2nd ed. Boston, MA: Lean Enterprise Institute, 2009.
    The Bahri Dental Group transformed their work and thinking from a traditional batch-and-queue approach to one focused directly on the needs of the patient

**Barnas, Kim. *Beyond Heroes: A Lean Management System for Healthcare*. Appleton, WI: ThedaCare Center for Healthcare Value, 2014.
    The dacare's story of developing a Lean Management System

Basic Understanding of Waste in a Non-Threatening Way Toast: Value Stream Mapping. Directed by Bruce Hamilton. DVD. Boston, MA: GBMP, 2009.
    Involving people to develop value stream maps and improve value streams

CEDAC: Cause & Effect Diagrams Adding Culture to Fosher Team Problem Solving. Directed by Bruce Hamilton. DVD. Boston, MA: GBMP, 2011.
    A simple, yet powerful, team-based, problem-solving technique

Change(over) is Good! Cut Costs and Increase Flexibility through Set Up Reduction. Directed by Bruce Hamilton. DVD. Boston, MA: GBMP, 2011.
    Basics of the changeover reduction process with real examples

Dennis, Pascal. *Lean Production Simplified: A Plain-Language Guide to the World's Most Powerful Production System*, 3rd ed. Portland, OR: Productivity Press, 2015.
    Basic handbook covering many Lean concepts

Ford, Henry. *Today and Tomorrow: Commemorative Edition of Ford's 1926 Classic*. Portland, OR: Productivity Press, 1988.
    Historical view of the roots of Lean. Winner of the 2003 Shingo Publication Award

Fukuda, Ryuji. *CEDAC: A Tool for Continuous Systematic Improvement*. Abingdon, England: Routledge, 2018.
    A how-to text focused on the technique known as Cause-and-Effect Diagrams Adding Cards (or CEDAC)

Fukuda, Ryuji. *Managerial Engineering: Techniques for Improving Quality and Productivity in the Workplace*. Portland, OR: Productivity Press, 1983.
    A guide and model for incremental implementation of best practices

Galsworth, Gwendolyn. *Visual Systems: Harnessing the Power of a Visual Workspace*. New York, NY: AMACOM, 1997.
Guide to creating a visual workplace where everything that needs to be understood can be seen at a glance

*Goldratt, Eliyahu M. *The Goal: A Business Graphic Novel*. Great Barrington, MA: North River Press Publishing Corp., 2017.
Graphic novel unlocks people's thinking about change

Go See: A Management Primer for Gemba Walks. Directed by Bruce Hamilton. DVD. Boston, MA: GBMP, 2013.
Introduction to a simple PDCA cycle for managers that will prepare them to unlock the brilliance of their team members and create a system and culture of initiative and accountability.

Graban, Mark. *Lean Hospitals: Improving Quality, Patient Safety, and Employee Engagement*, 2nd ed. Portland, OR: Productivity Press, 2011.
How to use Lean to improve safety, quality, access, and morale while reducing costs

Greif, Michael. *The Visual Factory: Building Participation through Shared Information*. Portland, OR: Productivity Press, 1991.
A guide to using visual control systems for daily management and improvement

Hamel, Mark R. *Kaizen Event Fieldbook: Foundation, Framework, and Standard Work for Effective Events*. Southfield, MI: Society of Manufacturing Engineers, 2009.
Addresses the root causes of poor kaizen event outcomes

Hamilton, Bruce and Pat Wardwell. *E2 Continuous Improvement System*. Newton, MA: GBMP, 2010.
Workbook on Lean philosophies, management systems, and tools/techniques

Hirano, Hiroyuki. *5 Pillars of the Visual Workplace: The Sourcebook for 5S Implementation*. Portland, OR: Productivity Press, 1995.
A source for implementing workplace organization, also known as 5S

*Hirano, Hiruyoki. *JIT Factory Revolution: A Pictorial Guide to Factory Design of the Future*. Portland, OR: Productivity Press, 1989.
Picture book of best Lean practices

Imai, Masaaki. *Kaizen: The Key to Japan's Competitive Success*. New York, NY: McGraw-Hill Education, 1986.
Good primer for implementing continuous improvement

*Imai, Masaaki. *Gemba Kaizen: A Commonsense Approach to a Continuous Improvement Strategy*, 2nd ed. New York, NY: McGraw-Hill Education, 2012.
Covers many aspects of Lean systems

Improvement Kata: Making Scientific Thinking a Skill for Life. Directed by Bruce Hamilton. DVD. Boston, MA: GBMP, 2014.
Improvement kata and coaching kata as a means for learning through behavior

*Japan Management Assoc., eds. *Kanban: Just-in Time at Toyota, Management Begins at the Workplace*. Translated by David J. Lu. Portland, OR: Productivity Press, 1986.
A fundamental resource on pull systems, written by the creator of the method

Liker, Jeffrey and Gary L. Convis. *The Toyota Way to Lean Leadership: Achieving and Sustaining Excellence through Leadership Development*. New York, NY: McGraw-Hill Education, 2011.
Leadership in a Lean environment

Miller, Lawrence M. *Healthcare Lean: The Team Guide to Continuous Improvement*. Annapolis, MD: Miller Management Press, 2012.
Fundamental tools and practices of working in teams to improve the quality of client service and eliminate waste

Mizuno, Sigeru. *Management for Quality Improvement: The 7 New QC Tools*. Portland, OR: Productivity Press, 1988.
  Fundamental coverage of graphical methods used to create and support strategic and statistical planning
Moments of Truth. Directed by Bruce Hamilton. DVD. Boston, MA: GBMP, 2009.
  Creating the right atmosphere to get ideas from employees
Monden, Yasuhiro. *Toyota Production System: An Integrated Approach to Just-In-Time*, 4th ed. Portland, OR: Productivity Press, 2012.
  Nuts and bolts of Lean methods
Monden, Yasuhiro. *Toyota Management System: Linking the Seven Key Functional Areas*. Portland, OR: Productivity Press, 1997.
  Describes a company-wide system model to TPS
*Ohno, Taiichi. *Toyota Production System: Beyond Large-Scale Production*. Translated by Andrew P. Dillon. New York, NY: Productivity Press, 1988.
  Fundamental philosophy and strategy behind Lean
Robinson, Alan G. *Ideas Are Free: How the Idea Revolution is Liberating People and Transforming Organizations*, 2nd ed. Oakland, CA: Berrett-Koehler Publishers, 2006.
  Discusses employee idea generation
Rother, Mike. *Toyota Kata: Managing People for Improvement, Adaptiveness and Superior Results*. New York, NY: McGraw-Hill Education, 2009.
  Describes improvement kata and coaching kata as a means of learning through behavior
Rother, Mike and John Shook. *Learning to See: Value Stream Mapping to Add Value and Eliminate Muda*. Boston, MA: Lean Enterprise Institute, 1999.
  How-to guide for value stream mapping
Sekine, Keniche. *One-Piece Flow: Cell Design for Transforming the Production Process*. Portland, OR: Productivity Press, 2005.
  A practical method for creating true single piece part and information flow
Sekine, Keniche. *Kaizen for Quick Changeover: Going Beyond SMED*. Portland, OR: Productivity Press. 1992.
  Good text for viewing changeover examples other than presses and machines
Shimbun, Nikkan Kogyo. *Poka-Yoke: Improving Product Quality by Preventing Defects*. Portland, OR: Productivity Press, 1989.
  Hundreds of examples of mistake-proof devices
*Shingo, Shigeo. *The Shingo Production Management System: Improving Process Functions*. Translated by Andrew P. Dillon. Portland, OR: Productivity Press, 1992.
Shingo, Shigeo. *Non-Stock Production: The Shingo System of Continuous Improvement*. Translated by Andrew P. Dillon. Cambridge, MA: Productivity Press, 1988.
  Overview of the methods and thinking of TPS
Shingo, Shigeo. *The Sayings of Shigeo Shingo: Key Strategies for Plant Improvement*. Translated by Andrew P. Dillon. Portland, OR: Productivity Press, 1987.
  Insight into the thinking behind Lean tools
*Shingo, Shigeo. *Zero Quality Control: Source Inspection and the Poka-Yoke System*. Translated by Andrew P. Dillon. Cambridge, MA: Productivity, 1986.
  Key source for developing the method of mistake-proofing
Shingo, Shigeo. *A Revolution in Manufacturing: The SMED System*. Translated by Andrew P. Dillon. Portland, OR: Productivity Press, 1985.
  Explanation of the practice of quick changeover to drastically reduce set-ups for any process or machine
Shook, John. *Managing to Learn: Using the A3 Management Process to Solve Problems, Gain Agreement, Mentor, and Lead*. Boston, MA: Lean Enterprise Institute, 2008.

Hands-on explanation of A3 thinking as a means to manage improvement

Shonberger, Richard J. *World Class Manufacturing: The Lessons of Simplicity Applied*. New York, NY: The Free Press, 2008.

Tools of Lean and reasons for success

Single Patient Flow: Applying Lean Principles in Healthcare. Directed by Bruce Hamilton. DVD. Boston, MA: GBMP, 2012.

The Bahri Dental Group transformed their work and their thinking from a traditional batch-and-queue approach to one focused directly on the needs of the patient, not on the needs of the practitioners.

Smith, Preston G, and Donald G. Reinertsen. *Developing Products in Half the Time: New Rules, New Tools*, 2nd ed. New York, NY: Wiley, 1997.

Best practices for development of new products, from concept to prototyping, design, testing and ramp-up

*Spear, Steve and Kent Bowen. "Decoding the DNA of the Toyota Production System." *Harvard Business Review* 77(5) (September 1999).

Description of the system that makes TPS a reality

Stalk, George and Thomas M. Hout. *Competing Against Time: How Time-Based Competition Is Reshaping Global Markets*. New York, NY: The Free Press, 2003.

An administrative view of the just-in-time (JIT) philosophy

Sugiyama, Tomo. *The Improvement Book: Creating the Problem-Free Workplace*. Portland, OR: Productivity Press, 1989.

A how-to handbook for total employee involvement

Sugiyama, Tomo and Japan Human Relations Association. *The Idea Book: Improvement through Total Employee Involvement*. Portland, OR: Productivity Press, 1988.

Good employee participation book

*Suzaki, Kyoshi. *New Manufacturing Challenge: Techniques for Continuous Improvement*. New York, NY: The Free Press, 1987.

Nuts and bolts of Lean methods

Suzue, Toshio and Akira Kohdate. *Variety Reduction Program: A Production Strategy for Product Diversification*. Portland, OR: Productivity Press, 1990.

Details engineering practices that can lead to significant reductions in pointless variety (i.e., variety not requested by the customer) in parts and processes

Toast Kaizen: An Introduction to Continuous Improvement & Lean Principles. Directed by Bruce Hamilton. DVD. Boston, MA: GBMP, 2005.

**Toussaint, John. *Management on the Mend: The Healthcare Executive Guide to System Transformation*. Appleton, WI: ThedaCare Center for Healthcare Value, 2015.

A playbook for healthcare leaders seeking to transform their organizations

Toussaint, John. *Potent Medicine: The Collaborative Cure for Healthcare*. Appleton, WI: ThedaCare Center for Healthcare Value, 2012.

A text for healthcare leaders or anyone who cares about improving the healthcare system in North America

**Toussaint, John and Roger Gerard. *On the Mend: Revolutionizing Healthcare to Save Lives and Transform the Industry*. Cambridge, MA: Lean Enterprise Institute, 2010.

The triumphs and stumbles of a seven-year journey to implement Lean in healthcare, which has slashed medical errors, improved patient outcomes, raised staff morale, and saved $27 million in costs without layoffs

Womack, James P. and Daniel T. Jones. *Lean Thinking: Banish Waste and Create Wealth in Your Corporation*, new ed. New York, NY: Simon & Schuster, Inc., 2003.

Modern-day review of the benefits of TPS

Worth, Judy, Tom Shukar, Beau Keyte, Karl Ohaus, Jim Luckman, David Verble, Kirk Paluska, and Todd Nickel. *Perfecting Patient Journeys: Improving Patient Safety, Quality, and Satisfaction While Building Problem-Solving Skills.* Cambridge, MA: Lean Enterprise Institute, 2012.
How to identify and select a problem, define a project scope, and create a shared understanding of what's occurring in the value stream

* Read or watch these general Lean references first.
** Read these healthcare references first.

About the Editors

Larry Anderson is a bronze, silver, and gold certified Lean practitioner (AME, Shingo, SME) with more than forty years of operational excellence experience. Larry is certified as a facilitator for all Shingo Institute workshops. He has a deep understanding of *Shingo Model* principles and application gained through more than eighteen years of work with the model. He has conducted more than twenty assessment site visits, many as the senior examiner/team lead and has performed numerous application desktop assessments. He has also participated in the Shingo publication review process and assisted two Shingo Institute Medallion recipients. Larry earned a BSME degree from Texas Tech University, is a registered professional engineer, serves as a Shingo representative on the Lean certification oversight and appeals committee, and is president of the AME Southwest region.

Dan Fleming, director of consulting services at GBMP, is a Shingo Prize recipient and a certified Shingo Institute workshop facilitator. He is Lean silver certified and brings more than thirty years of experience in operations and engineering to GBMP, including more than twenty-five years of hands-on experience learning, leading, and teaching the principles and tools of the Toyota Production Systems and continuous improvement. He was the lead developer of GBMP's highly regarded Lean in Healthcare Certificate course, a comprehensive program that was one of the first of its kind in healthcare. Dan has been a valued speaker at regional and national conferences. Prior to becoming director of consulting services, Dan was a continuous improvement manager at GBMP for fourteen years, working with a wide range of organizations including healthcare, electronics, medical device, pharmaceutical, equipment manufacturers, food processing, machine shops, contract manufacturers, and warehouse and distribution. Dan holds a BS in electrical engineering technology from Northeastern University.

Bruce Hamilton, president of GBMP, is also a director emeritus for the Shingo Institute, a senior Prize examiner, and a certified Shingo Institute workshop facilitator. Bruce is a past recipient of the Shingo Prize in both the business and academic categories and is an inductee into the Shingo Academy (with five awards in all). In 2015, he was also inducted into the AME Manufacturing Hall of Fame. A sought-after speaker on management's role in Lean transformation, Bruce's clients have included Raytheon, Beth Israel Deaconess Hospital, OC Tanner, and New Balance, as well as many smaller organizations. He is the creator of the well-known *Toast Kaizen* video, as well as numerous other award-winning Lean training videos including *Single Patient Flow* and, most recently, *Improvement Kata*. Bruce is co-author of *e2 Continuous Improvement System*, a comprehensive learn-by-doing guide for Lean transformation. He also publishes the blog, *OldLeanDude*, which is about understanding the Toyota Production System and gaining its full benefits, and he hosts the free monthly webinar, "Tea Time with the Toast Dude." Before joining GBMP, Bruce held management positions in marketing, IT, operations management, and general management, and, in 1990, he led his organization to a Shingo Prize. He is equally at home in administrative, operational, and healthcare environments. As an early adopter of Lean, his factory was visited by Shigeo Shingo, Shigihiro Nakamura, and Ryuji Fukuda, among others. From 1994 to 1998, he was coached by Hajime Ohba and TSSC. Bruce holds a BA and attended Bowdoin College and the University of Arizona.

Pat Wardwell is a bronze, silver, and gold certified Lean practitioner (AME, Shingo, SME) with more than thirty years of operational excellence experience. She is a certified Shingo Institute workshop facilitator as well as a Shingo Prize examiner. In 2011, Pat was a co-recipient of a Shingo Prize in the research and professional publication category. She is also the co-author of *e2 Continuous Improvement System*, a comprehensive learn-by-doing guide for Lean transformation. Pat has served in a variety of operational and management roles, including vice president of operations at a company that was a 1990 Shingo Prize recipient. Involved with continuous improvement efforts on the shop floor and in support areas since 1987, Pat has trained and coached many companies, some of whom have received recognition through the Shingo Prize process. Pat earned a BA from the University of Maine and an MBA from Bentley University. She serves as an AME representative on the Lean certification oversight and appeals committee, and is a lead examiner and Awards Council member for the AME Excellence Award and long-time AME volunteer.

Index

Note: Page numbers in *italics* indicate a figure on the corresponding page.

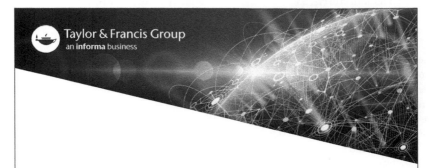